D0984059

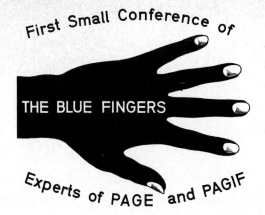

First Small Conference of

THE BLUE FINGERS

Experts of PAGE and PAGIF

Electrophoresis and Isoelectric Focusing in Polyacrylamide Gel

K. Abraham

R. C. Allen

N. Catsimpoolas

A. Chrambach

W. Dames

K. Felgenhauer

W. Giebel

D. Graesslin

A. Griffith

U. Grossbach

H. Haglund

H. Hoffmeister

H. Hunter

T. Jovin

G. Kapadia

O. Kling

H. R. Maurer

C. J. O. R. Morris

S. Nees

V. Neuhoff

G. Philipps

P. Pogacar

E. G. Richards

R. Rüchel

D. Rodbard

H. Stegemann

L. Strauch

J. Uriel

G. Utermann

O. Vesterberg

Electrophoresis and Isoelectric Focusing in Polyacrylamide Gel

Advances of Methods and Theories, Biochemical and Clinical Applications

Editors
R.C. Allen · H.R. Maurer

Walter de Gruyter · Berlin · New York 1974

Proceedings of the Small Conference held at Tübingen, Germany,
on October 6–7, 1972

Editors:

Robert C. Allen, Ph. D.
Associate Professor of Pathology at the Medical University of South Carolina,
80 Barre Street,
Charleston, South Carolina 29401, USA.

Professor Dr. *H. Rainer Maurer,*
Max-Planck-Institut für Virusforschung, Abteilung für Physikalische Biologie,
D-74 Tübingen, Spemannstr. 35, Germany.

Present address:
Pharmazeutisches Institut der FU, D-1 Berlin 33, Königin-Luise-Straße 2–4,
Germany.

This volume contains 115 figures and 19 tables.

ISBN 3 11 004344 0
Library of Congress Catalog Card Number 73-94225

Table of Contents

Chapter 6. Preparative Methods

Chapter 7. Micro Methods

Chapter 8. Biochemical Applications

Chapter 9. Clinical Applications

Introduction. Welcome and Introductory Remarks

H. Rainer Maurer

Ladies and Gentlemen,

It is a great pleasure for me to welcome you all to the first Small Conference of the Blue Fingers at Tübingen. Modern biology, biochemistry and clinical chemistry widely use the techniques of the Blue Fingers and need their experiences. But who are the Blue Fingers and why are they assembled here? When I showed this hand to a number of colleagues, I soon realized that this figure is an excellent way to distinguish the lab bench experts from those knowing the techniques just by hearsay. May I suggest that you find out the distinction yourself. It is the aim of this conference to provide an extensive exchange of views and ideas among the Blue Fingers and to tell the White Fingers how not to blue their fingers.

Anyhow, Dr. Allen and I hope that we will determine the requirements for the standardization of PAGE and PAGIF. This is a prerequisite for routine work in biochemistry and clinical chemistry. Moreover, we expect to hear about theoretical backgrounds and newly devised methods. Big science can only grow on safe technical grounds. But one must know how to do it. And, therefore, we are here in order to learn more about our colleagues' experiences.

Before stepping away, Dr. Allen and I wish to thank those who made this conference possible, in particular the Erwin-Riesch-Stifung, the European Molecular Biology Organization and the exhibiting firms. Last not least we gratefully acknowledge the assistance in the organization of the meeting by Mrs. G. Pauldrach, Miss G. Pauldrach, Mr. R. Witrofsky, Miss A. Weissmann and my wife.

I have two final remarks. May I ask the speakers to keep on time to allow sufficient time for discussions. And to my German colleagues: You are welcome to speak in German if you prefer.

Now may I ask Dr. Chrambach to chair the first session on principles and new aspects of polyacrylamide gel electrophoresis.

Chapter 1. Polyacrylamide Gel Electrophoresis: Physico-chemical Properties of the Gel

1.1 Physico-chemical Measurements in Poly-acrylamide Gels

C. J. O. R. Morris and *P. Morris*

In 1966 Morris published the results of a preliminary study of the relation between the electrophoretic mobilities of some model proteins in polyacrylamide gels and the composition of the gels, expressed in terms of the total monomer content T, and the % cross-linking C. These results, in contrast with those of some earlier investigators, supported a relation of the type:

$$\log M = \log M_0 - K\,T \tag{1}$$

or

$$\log M/M_0 = -K\,T = M' \tag{1a}$$

where M is the measured mobility ($cm^2/s/V \times 10^5$), M_0 is the extrapolated value of M when T = 0, and K is a constant. I suggested the term reduced mobility for the dimensionless ratio M'.

Further comparison of these results with earlier experiments (Fawcett and Morris, 1966), using molecular sieve chromatography on granulated polyacrylamide gels, showed that the reduced mobility M' was a linear function of the chromatographic partition coefficient K_{av}, so that:

$$K_{av} = a\,M' + b \tag{2}$$

where a and b are constants, a having a value near one, and b being small and negative, and apparently a function of the cross-linking C. These results provided for the first time a unified theory of molecular sieve chromatography and electrophoresis in gels. It must however be emphasized that the evidence for both equations (1) and (2) is wholly experimental, and their validity is entirely dependent on the quality of that evidence. It was pointed out in this paper (1966) that equation (2) follows if the space available for solute molecules is similar in chromatography and electrophoresis, while Rodbard and Chrambach (1970) have subsequently given a more general derivation.

Equation (1) can be derived from equation (2) on the basis of several of the current theories of molecular sieve chromatography (Ogston 1958; Laurent and Killander 1964; Giddings et al. 1968; Hjerten 1971).

In 1971 Morris and Morris published the results of a detailed investigation which we believe provides unequivocal support for the unified theory. As this Symposium is concerned with methodology, we thought it might be of general interest to describe in some detail our improved technique for measurement of the electrophoretic mobilities of proteins in polyacrylamide gels.

In contrast to the earlier experiments we decided to carry out both chromatographic and electrophoretic experiments in a single medium, Tris-HCl buffer, pH 8.87, I = 0.05. The use of this relatively high ionic strength was necessitated by the need to minimize protein-gel interactions in the chromatographic experiments. Under these conditions very efficient cooling was required during electrophoresis.

As in the earlier series, we decided to measure absolute electrophoretic mobilities, mainly because we anticipated that this series of experiments would occupy 2–3 years, and we therefore required a technique which could be readily checked at intervals during this period. We also observed in some preliminary experiments that the temperature coefficients of mobility of proteins, particularly basic proteins, can vary widely. This observation, which does not appear to have been widely reported in the literature, makes all relative mobility measurements suspect at the 2–3 % precision level, unless extreme precautions are taken to ensure temperature uniformity within the gel. The model proteins and any marker ions used should also have very similar temperature coefficients of mobility.

The requirement that the experimental conditions should be compatible with both the electrophoretic and chromatographic experiments made it essential to use gels which were fully equilibrated with the buffer medium. In any case we consider this to be essential for any experiments in which a knowledge of the true gel composition is required, as with weakly cross-linked gels considerable solvent uptake can occur. Figure 1 (after Morris and Morris 1971) shows the water regains of our gels in 0.25 M Tris – 0.05 M HCl(Data in Table 1). The broken line corresponds to the water content of the monomer solutions before polymerization, so that if polymerization is complete, and there is no volume change.in the reaction, gels with T-values on the line will neither swell or shrink. It is evident that 1, 3 and 5 % C-gels all swell to varying extents, while 7 and 9 (and also 11) % C-gels show negligible solvent uptake. It must be emphasized that these data are only valid for the particular buffer medium used, and solutions with lower osmolalities may show substantially

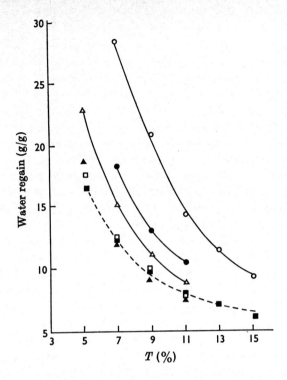

Fig. 1. Water regains (g of water/g of dry polyacrylamide) of polyacrylamide gels.
○, X1; ●, X3; △, X5; □, X7; ▲, X9; ■, theory.

Table 1. Water regains of polyacrylamide gels from 0.25 M tris — 0.05 M HCl buffer,
pH 8.87. (G water / G dry gel)

T \ C	1	3	5	7	9
5			23.0	17.8	19.1
7	28.6	18.45	14.6	12.8	12.0
9	21.0	13.05	10.5	10.3	9.1
11	14.4	10.5	8.9	7.9	7.5
13	11.7				
15	9.65				

greater swelling. In fact 1 % C-gels may never come into osmotic equilibrium with water alone.

Since the chemical aspects of acrylamide polymerization are to be treated in a subsequent paper, we will not consider them further here, except to summarise the experimental conditions which we have found to give reproducible gels over a period of several years.

1. Use of riboflavin as photochemical initiator, and tetramethylethylene-diamine (TEMED) as basic catalyst in order to eliminate the possibility of incorporation of ionic groups into the gel matrix.
2. Gel polymerization is carried out in water rather than buffer.
3. Efficient removal of the heat liberated during polymerization. The reaction is exothermic with $\Delta H = - 84$ kJ per mol, so that this is particularly important with high T gels.
4. Polymerization is carried out at $30°$ C for 20 h with continuous irradiation to ensure completion of the reaction.
5. Under these conditions the absolute purity of the monomers appears to be relatively unimportant, so that recrystallized materials offer no special advantages.

When polymerization was complete, the 45 X 12 X 0.5 cm gel sheets were removed from the moulds, and subjected to a cycle of intensive washing with water, 0.5 M NaCl, water, and finally Tris-HCl buffer until the supernatant liquid showed no change in pH, conductance or ultra-violet absorption between 2 successive 2 hour washes. The gel sheets were stored in buffer and always used within 14 days of polymerization.

The apparatus used for measurement of mobilities is shown in figure 2. Although simple in concept, it incorporates many detail refinements which contribute to its efficiency. It consists essentially of a 40 X 12 X 0.5 cm (nominal) horizontal gel sheet G, dipping into anode and cathode vessels A and C, and in contact with the upper and lower cooling plates P_U and P_L maintained at $10 \pm 0.2°$ C by coolant circulated from an external thermostat.

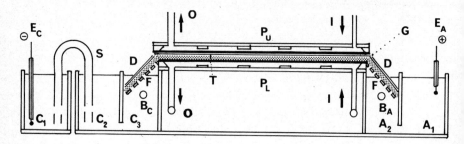

Fig. 2

P_U and P_L have internal glass surfaces to take advantage of the superior thermal conductivity of glass (10–15 times that of plastics). The inlet tubes I, I (and also the outlet tubes 0, 0) are in parallel to eliminate thermal differences between the cooling plates, and the thermal flow I to 0 opposes the direction of electrophoretic migration E_C to E_A. I, I are smaller than 0, 0 to prevent pressure distortion of the plane glass plates. The half-thickness baffles inside the cooling plates promote turbulence, and direct the turbulent flow on to the glass plates. This arrangement would dissipate 50 W from a 5 mm thick gel at an ambient temperature of 10^0 C with a temperature increase of less than 0.5^0 C within the gel. Actual operating conditions were kept below 25 W.

Our first design, following common practice, used thick filter-paper wicks to connect the gel to the electrode vessels. This arrangement occasionally gave erratic results due to inhomogeneous electric fields. These were entirely eliminated by the design shown here, in which the gel slab G dips directly into the buffer reservoirs, supported on the perforated grids F, F, and protected against evaporation by the paper and polyethylene covers D, D. This arrangement presumably reduces both contact irregularities and Bethe-Toropoff effects at the gel — buffer interfaces.

Operation at relatively high currents for extended periods will result in considerable changes in pH and conductance near the electrodes, and these must be considered in the design of the electrode vessels in order to prevent such changes reaching the vicinity of the gel. Our design is based on the fundamental observations of Svensson (1955), who showed that in general the acids liberated from a salt by electrolysis at an anode were less dense than the salt solution itself, and would consequently float upwards, while the corresponding bases liberated at the cathode were denser than the salt solution, and tended to sink to the bottom of the electrode vessel.

In our design the anode E_A is placed near the top of the vessel A_1, so as to minimize the effects of stirring by gas evolution from the electrode, and also to ensure that the pH and conductance in the compartment A_2 will not change until the whole of the buffer in A_1 is used up and the lighter low — pH boundary passes under the partition between A_1 and A_2. Conversely at the cathode E_C, no pH or conductance change from C_1 can enter C_2 until the buffer in C_1 is used up and the denser catholyte can pass over the four parallel siphons S connecting C_1 to C_2. Gas evolution at E_C actively promotes mixing in C_1 and prevents denser solution accumulating at the bottom. Coolant at 10^0 C circulated through the horizontal cooling tubes B_A, B_C helps to maintain more uniform temperatures in compartments A_2 and C_3, and if necessary to promote downward thermal convection in these compartments. A total volume of 5 L of buffer was employed in this electrode system. No changes in pH, conductance or temperature were observed in A_2 or A_3 after 20 h at up to 25 W.

The apparatus was operated from a constant-current regulated power supply. Since the temperature and ionic environment of the gel are kept constant, its electrical resistance R will also be constant, so that the iR drop across it, and hence the potential gradient will be kept constant, irrespective of any conductance changes in A_1 and C_1. In practice the potential gradient measured directly across the upper surface of the gel varied by less than $\pm 1\%$ during 20 h.

Ten protein samples were introduced into the gel on filter paper strips inserted into slits cut in the gel at T. The validity of the method was monitored by migration of coloured proteins through empty slots cut in an experimental gel. The method has the advantage of introducing the solutes into a part of the gel remote from its ends.

At the end of the experiment, when the slowest model protein, carbonic anhydrase, had migrated at least 30 mm, the gel was removed, and stained with Naphthalene Black in 7% acetic acid at 100° C. Destaining was carried out electrolytically. The use of hemoglobin as a marker protein enabled the absence of dimensional changes in the gel during staining and destaining to be checked directly.

Under these conditions the distances traversed by the test proteins were linear both in time and potential gradient, conditions which we regard as indispensable for meaningful measurements of electrophoretic mobility. Figure 3 shows the

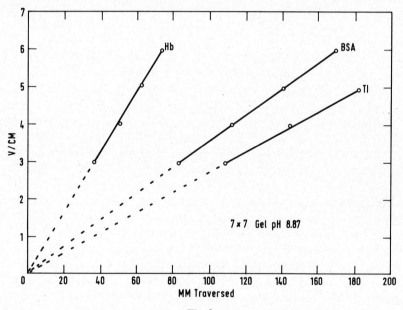

Fig. 3

migration distance — potential gradient data for three proteins, soyabean trypsin inhibitor (TI), bovine serum albumin (BSA) and human haemoglobin (Hb) in a 7 × 7 gel. We have similar data for the other gel compositions.

As a result of some 450 individual measurements of the mobilities of the eight model proteins in the 20 gel compositions studied (σ = 0.05 mobility units, mean coefficient of variation ± 2.0%), we are now satisfied that the logarithmic equation (1) represents the relation of electrophoretic mobility M to monomer content T with high precision.

Comparison with the parallel chromatographic experiments also confirmed equation (1) and enabled the values of the constants a and b to be determined with adequate precision. The experimental values are given in Table 2. We may therefore conclude that both separation methods operate by a very similar molecular sieving mechanism, and the way is now open to consider further the physical meaning of these constants.

Table 2. Constants of the equation K_{av} = a M' + b

C	1	3	5	7	9
a	1.21	1.15	1.04	1.07	1.17
b	−0.24	−0.20	−0.07	−0.07	−0.07

A possibility which has not yet been wholly excluded is the intervention of electro-osmotic effects, and although these effects are undoubtedly very small in polyacrylamide gels, the magnitudes of the constants a and b are such that an explanation on these lines is possible. We are now actively considering alternative methods for the measurement of very small electro-osmotic flows in gel media.

References

Fawcett, J. S. and C. J. O. R. Morris, Separation Sci. *1*, 9 (1966).
Giddings, J. C. et al. J. phys. Chem. *72*, 4397 (1968).
Hjerten, S., J. Chromat. *50*, 189 (1970).
Morris, C. J. O. R., Protides of the Biological Fluids, ed. H. Peeters, *14*, 543 (1966).
Morris, C. J. O. R. and P. Morris, Biochem. J. *124*, 517 (1971).
Ogston, A. G. , Trans. Faraday Soc. *54*, 1754 (1958).
Rodbard, D. and A. Chrambach, Proc. Nat. Acad. Sci. USA *65*, 970 (1970).
Svensson, H., Acta chem. scand. *9*, 1689 (1955).

1.2 Polymerization Kinetics and Properties of Poly-acrylamide Gels

E. G. Richards and *R. Lecanidou*

Introduction

Polyacrylamide gels are prepared by polymerising an aqueous solution of acrylamide and BIS (N, N′ methylene bisacrylamide) using a free radical generating catalyst system. The reaction proceeds by a vinyl addition polymerization process as described by Flory (1).

The resulting gels show a continuous spectrum of properties from being soft, almost viscous, transparent gels which swell greatly in excess sovent on the one hand to hard, brittle, ringing gels that are opaque and do not swell on the other. The exact properties of a particular gel depend on the total weight concentration of monmer (T), the proportion by weight (C) of the cross linking reagent and to some extent on the concentration of the catalyst added to induce polymerisation.

Kinetics of polymerisation

The polymerization kinetics of gels themselves are not easily studied; instead we have investigated the polymerization of solutions of acrylamide alone, which form viscous solutions rather than gels, by following the viscosity of the reaction mixture with time. We express our results in terms of the specific viscosity of the mixture and this parameter is expected to increase linearly with the conversion of monomer to polymer. A typical reaction curve is shown in figure 1. It is characterised by an induction period, t_s, followed by a sharp rise in viscosity which eventually levels off to a constant value, η^{max}. In so far as the conversion of monomer to polymer is a first order process (1), t_r is a measure of the rate constant. η^{max} is related to the mean molecular weight of the polymer formed.

Our experiments have shown that t_s is proportional to $I^{-3/2}$ where $I = \sqrt{(SD)}$ and S is the concentration of ammonium persulfate and D that of TEMED added as catalyst. t_s also depends on the amount of inhibitor present which may be dissolved oxygen or substances introduced to delay the onset of polymerisation.

In later experiments with gels, the setting time was found to be related to the catalyst concentrations in the same way.

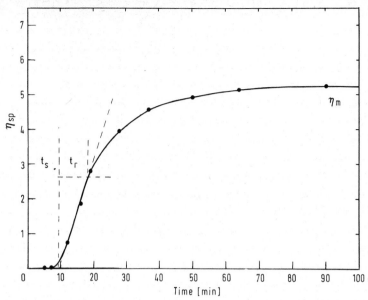

Fig. 1. Time course of polymerisation of 2.5 % acrylamide in water saturated with O_2 at 25°. The reaction was initiated by 6.6 mM ammonium persulphate and 6.9 mM TEMED.

In accordance with the kinetic scheme for vinyl polymerisation discussed by Flory (1), t_r was found to be inversely proportional to I and η^{max} was found to decrease as I increased. We also found that η^{max} increased exponentially with the acrylamide concentration.

These and similar observations led to the following conclusions of practical relevance:

1. Excess catalyst leads to short polymer chains. This is likely to be true also in the case of gels. Since the formation of a gel depends on the production of closed rings of polymer chain linked by BIS residues, a gel cannot form if the mean chain length is a lot less than the number of acrylamide residues separating the BIS residues (as determined by C). In accord it was regularly observed that gelation did not take place in the presence of excess catalyst.

2. If too little catalyst is employed the induction period (t_s) is extended and unless air is rigorously excluded, gelation my never occur.

3. Excess ammonium persulfate alone can inhibit the formation of gels. The explanation of this effect is not clear but the practical consequence is that it is best to use equal molar concentration of ammonium persulfate and TEMED.

4. There is an optimum catalyst concentration which depends on the monomer concentration (T), the presence of inhibitors and how long one requires to wait before the gel sets. Catalyst concentrations in the range 1−10 mM are usually optimum.

Changes on polymerisation

The polymerization reaction is exothermic and the resulting gel occupies
less volume than the original monomer mixture. The latter effect can give
rise to distortions at the top of the gel and to a tendency of the gel to break
away from the walls of the containing vessel. These effects are more pronounced at high gel concentrations, but we have not managed to devise a
remedy. Both these effects also give rise to inhomogineities in the gel. This
effect can be mitigated by slowly rotating the vessel containing the reaction
mixture about a horizontal axis while polymerization is taking place. In this
way gels which are nearly optically homogeneous can be prepared (2).

Structure of gels

It is assumed that a polyacrylamide gel consists of long chains of polyacrylamide cross-linked at intervals to form a random three dimensional net. We
may refer to the crosslinks as knots and the chains linking them together as
links. It is useful to consider an 'ideal' gel in which the knots are distributed
randomly throughout the volume. In this case the mean distance between a
knot and its nearest neighbor is well defined and depends on the concentration
of knots only. We also suppose that the knots are connected by 'random coil'
links between nearest neighbour knots. The r.m.s end to end distance of such
a coil is related to the number of residues in the chain and hence (if all
acrylamide residues are incorporated in chains) to T the total concentration.

If these two dimensions are to be equal there must be a unique relation between
T and C — such defines the ideal gel.

If the concentration of knots is greater than implied by this relation it is likely
that they must cluster together (3, 4); similarly if there are too few knots the
r.m. s. end to end distance of the links will be less than the equilibrium value.
This situation would be remedied if the gel were to swell. Gels made with
values of T and C corresponding to these conditions may be referred to as
clustered and crumpled gels respectively (3). It is clear that this model of gel
structure is drastically oversimplified; we have for instance ignored the
possibility of entanglements (which will act like extra crosslinks) formed by
two linked rings; links between non-nearest neighbor knots, etc. Nevertheless
this simple model permits a qualitative understanding of some properties of gels.

Swelling

When a gel is immersed in excess water it will generally absorb solvent and
swell (by a volume factor Q). The swelling of crosslinked polymers has been
much studied in relation to rubber technology and proceeds till a further

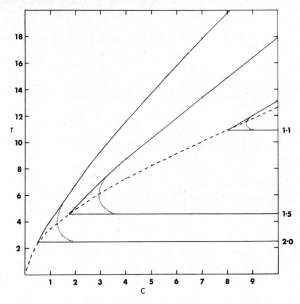

Fig. 2. Theoretical contours of constant swelling factor.

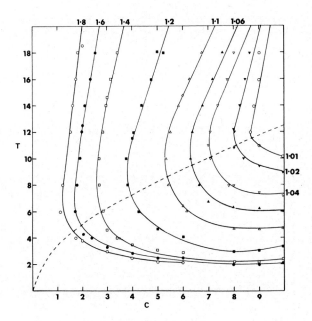

Fig. 3. Experimental contours of constant swelling.

decrease in the free energy by dilution concommitant on swelling is balanced
by an increase on stretching the links. Theories of gel swelling along these
lines (1) permit an estimation of the equilibrium value of Q in terms of the
initial values of T and C (3). However if the gel starts off by being 'crumpled'
it will first tend to swell till the new values of T fit the relation with C
corresponding to an ideal gel, and then continue to swell in the normal manner.
These considerations lead to a prediction of the value of Q corresponding to
various pairs of values of T and C as shown in figure 2 in the form of contours
of constant Q. The shape of these theoretical contours bear a resemblance to
experimental contours as shown in figure 3.

Theories of rubber elasticity suggest a relation between swelling and elasticity.
We have not studied quantitatively the elastic properties of gels but have noted
that gels that swell a lot are soft and easily deformable whilst gels that swell
little are rigid in agreement with these theories.

Turbidity

It is well known that polyacrylamide gels are rarely optically clear and may
in extreme cases be milky to the point of opacity. The turbidity (measured by
the absorbance at 550 nm) shows an inverse relation to the swelling factor,
Q, as shown in figure 4. It seems likely that turbidity is related to concentration variations on a micro-scale and thus to the presence of clusters of knots.

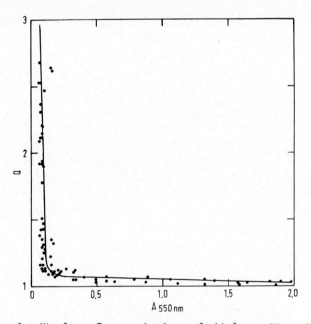

Fig. 4. Plot of swelling factor, Q, versus absorbance of gel before swelling at 550 nm.

Excluded volume

An attractive theory of gel permeation chromatography (6) relates the K_d for a macromolecule to the fraction, a, of the gel volume that can be occupied by the molecule. Ogston (5) has calculated the relation between a and the concentration for a random array of rods. The resulting expression quantitatively explains the behaviour of proteins in gel permeation chromatography in polyacrylamide gels (4). There is also evidence that the ratio of the mobility of a macromolecule in a gel to that in free solution is also related to a (7, 8, 9). Nevertheless the Ogston theory makes no mention of the crosslinks. We have found experimentally for 5S RNA that the mobility passes through a minimum as the value of C increases as shown in figure 5. Other authors have reported the same effect for proteins (7, 10).

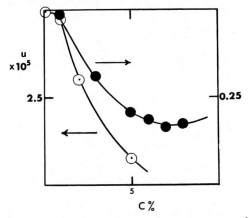

Fig. 5. Mobilities of renatured 5S RNA in 0.05 M tris-HCl pH 7.8 at 25^0 versus C, the proportion by weight of BIS in the monomer mixture. The total monomer concentration was kept constant at 12.5% (lower curve) or 18% (upper curve).

We suggest that the local concentration of gel material is greater in the vicinity of a knot and that these concentrations further reduce the volume available to a macromolecule. On this basis we expect the volume fraction available to decrease as C increases. As the concentration of knots further increases we we have argued that they start to cluster. This may result in the gel surrounding the clusters to become less concentrated and a net increase in the volume available, in accord with experiment.

Conclusion

The theories outlined above allow a qualitative understanding of most of the relevant properties of polyacrylamide gels and their dependence on their composition. From a practical point of view the best compromise between turbidity and swelling is achieved when the proportion of crosslinks, C, is 5%. In these circumstances gels of all useful concentrations, T, are almost optically clear and swelling is not troublesome except at the very lowest concentrations. Moreover, at this value of C the mobility ratio is close to its minimum.

References

(1)   Flory, P. J., Principles of Polymer Chemistry, Cornell University Press, New York, 1953.
(2)   Elliott, A., Personal communication.
(3)   Richards, E. G. and C. J. Temple, Nature Physical Sciences *230*, 92 (1971).
(4)   Fawcett, J. S. and C. J. O. R. Morris, Separation Science *1*, 9 (1966).
(5)   Ogston, A. G., Trans. Faraday Soc. *54*, 1754 (1958).
(6)   Laurent, T. C. and J. Killander, J. Chromatog. *14*, 317 (1964).
(7)   Morris, C. J. O. R. and P. Morris, Biochem. J. *124*, 517 (1971).
(8)   Rodbard, D. and A. Chrambach, Proc. U. S. Nat. Acad. Sci. *65*, 970 (1970).
(9)   Richards, E. G. and R. Lecanidou, Anal. Biochem. *40*, 43 (1971).
(10) Hjerten, S., S. Jerstedt and A. Tiselius, Anal. Biochem. *27*, 108 (1969).

Chapter 2. Polyacrylamide Gel Electrophoresis: Theory and Practice of Optimization and Standardization

2.1 Requirements of Reproducible and Standardizable Polyacrylamide Gel Electrophoresis

H. Rainer Maurer

One major reason why Dr. Allen and I organized this small conference was to find a satisfying answer to the intriguing question: Why is polyacrylamide gel electrophoresis (PAGE) relatively little recognized and utilized in practical clinical chemistry *on a routine basis,* despite its widespread use in biochemical and biological research? I have attempted to compile several factors which, I think, are responsible for the situation up to date and which I wish to discuss with you (1–3). I hope we can resume this discussion at the end of the conference in order to find out what we really need to do in the near future.

I think we can distinguish two major classes of factors which may be termed (1) *PAGE Technology* and (2) *Clinical Pattern Evaluation.* Let me discuss the first class of factors now:

Sample collection, preparation and handling

The samples should be collected and stored in closed small vials under sterile conditions. Serum may be kept at room temperature for a few days without significant change of most proteins except particular lipoproteins and α_1-anti-trypsin. For longer storage, it may be frozen in small aliquots to avoid many thawings. Freezing should be done rapidly to prevent protein aggregation. Preservatives like sodium azide or merthiolate may be added.

Apparatus operation

Obviously, each apparatus has its advantages and drawbacks. However, the necessity to analyse a large number of samples simultaneously on a routine basis should favour the choice of an apparatus which minimizes the risk of apparatus-inherent problems. Devices producing *slab gels* offer several advantages over those designed for cylindrical gels.

1. Many samples can be electrophoresed under completely identical conditions in a single gel. Thus, the patterns are directly comparable.

2. Joule heating can be better dissipated from 1–3 mm thick flat gels as normally used.

3. Due to their rectangular cross-section flat gels are easily evaluated by densitometry and photography without the risk of optical artifacts.

4. Reference samples may be included with unknowns in the same gel for exact side-by-side comparison.

5. More uniform gels may be cast in flat dishes which in turn improves the reproducibility.

6. Flat gels may be easily combined with other separation techniques for two-dimensional analyses.

7. Flat gels are much easier to dry for storage and autoradiography.

8. Less time is needed for the preparation of gels for multisample analysis.

Gel formation

Several aspects of gel formation have recently been discussed by Chrambach and Rodbard (4). Let me just mention the following points:

1. The gel components acrylamide, N,N'-methylenebisacrylamide and tetramethylethylenediamine should be as pure as possible.

2. Type and quantity of polymerization catalysts should be selected such that polymerization is completed within about 20 min and monomer to polymer conversion should reach at least 95%.

3. Inhibitors of polymerization should be avoided. Among these, atmospheric oxygen is the most significant. At contact points of different apparatus materials (glass – plexiglass – rubber) polymerization may be inhibited.

4. Since the polymerization rate is highly temperature-dependant, it is necessary to control the *temperature* to achieve reproducible gels. Moreover, the exothermic reaction of the polymerization may change the gel dimensions: it may lead to swelling or contraction. Therefore, the temperature during polymerization should approach that of the following electrophoresis.

5. Pre-electrophoresis may, in few cases, be necessary to eliminate catalyst artifacts due to the reaction between sample and catalyst.

In any case the gels should be tested with known standards to confirm their separation capacity. The best way is to co-electrophorese reference molecules of known size, shape, molecular weight and mobility under identical conditions.

Electrophoretic conditions

These comprise the usual parameters required for reproducible polyacrylamide gel electrophoresis:

1. Gel concentration or density expressed by the total acrylamide (T) and Bis (C) content.

2. Buffer composition, in addition indicating ionic strength and pH at the particular temperature.

3. Electric conditions minimizing overheating of the system.

4. Electrophoresis time.

Table 1. Factors influencing standardization and reproducibility of polyacrylamide gel electrophoresis (PAGE)

PAGE Technology		Clinical Pattern Evaluation	
Factor	Evaluation	Factor	Evaluation
Sample collection		Genetics	
preparation		Sex	
handling (storage)	2	Patient age	
Apparatus operation	1	Pre-diseases	
Gel formation	2	Stage of disease	
purity of gel components		Pluri-disease	3
polymerization rate and			
temperature		Diet	
catalyst type, quantity		Therapy	
and inhibitors			
Electrophoretic parameters	2	Diurnal and seasonal rhythms	
gel concentration (density)		Stress	
buffer system			
electric conditions			
Quantitative evaluation	2		
fixation			
staining			
photography			
densitometry			
computerized data output			

Evaluation of informations available so far: Sufficient (1), more data needed (2), insufficient (3)

Quantitative evaluation of the separation patterns

Following electrophoresis the separated sample components have somehow
to be detected and evaluated. At present the reproducible quantification of
gel electropherograms presents a major problem, not only between laboratories because of the lack of standardized procedures, but also due to the
instruments available for quantitative evaluation by densitometry and photography. These problems will be treated in more detail by Dr. Allen in this
volume. Let me just remind you that fixation and staining of proteins and
peptides are subject to conditions which may render them irreproducible.
For example, the dye-protein bond is, in most cases, not a covalent linkage
and may, therefore, be easily cleaved by extensive washing and light exposure.
Controlled conditions are hence extremely important. Moreover, the dye-
binding capacity of the different protein species within a given sample may
vary considerably, which means that only relative values can be obtained.
Nevertheless, since the molar extinction coefficients of the most commonly
used dyes are higher than those of the sample molecules (in the order of
50 000) *staining* is still the most sensitive detection method using absorption
spectrometry.

It should be stressed that *fixation* within the gel is at least as critical as staining. Irreversible fixation of small molecules down to 1000 MW is difficult to
achieve. Since the fixation process is poorly understood on a molecular level,
it seems to me that we need more information about this.

It should also be noted that fixation and staining can only detect *mass
differences* of proteins between different samples. Mass differences of proteins
may be a relatively late event in the course of a disease. It would therefore be
desirable to have methods at hand capable to detect subtle changes specific
for a particular disease as early as possible. The determination of turnover
rates normally requires the use of radioisotopes. Such studies are rarely performed because of expected radiation damages. However, calculations show
that the radiation load by ^{3}H may often be much lower than that by any
quick X-ray examination. Also certain enzymes may reveal different activities
much earlier than other proteins. It follows from these considerations that
the main problem will be to find the most indicative proteins.

Now let me briefly discuss the second class of factors with the heading
Clinical Pattern Evaluation. High resolution polyacrylamide gel electrophoresis reveals 5—10-fold the number of proteins found in human serum by other
techniques, such as cellulose-acetate electrophoresis or ultracentrifugation.
This has given rise to some confusion and error in the assessment of the
potentialities of the method. In many cases it has become difficult to inter-
prete the complicated patterns, in order to correlate them with specific

diseases. Why? The method is capable of detecting pattern differences of healthy donors due to differences in their genetics, sex, stage of disease, age and diet, as listed in the table. The many interrelationships possible between these factors make it almost impossible to draw any acceptable conclusions. What can we do in this situation? It appears to me that one first approach is to adopt a new rationale of protein classification. Available data on the many blood proteins should permit a satisfactory identification and classification (3). Then a standardization of the listed factors may become necessary. For example, specific correlations between human genetics and genetically controlled proteins, such as the haptoglobins and the Gc proteins should be established. The final step would be the search for possible correlations between specific diseases and distinct protein patterns. Here quantitative differences may become relevant before qualitative differences are obvious. Consequently, predisease stages, for example, may be diagnosed quantitatively rather than qualitatively. Apparently, this is an enormous task forcing us to limit the goal and to concentrate our interests onto specific classes of proteins promising maximum information. Moreover, this task must also induce the clinician to refine his way of questioning. This means that we may have to take all these factors into account. It appears to me that we know very little about them up to now.

Consequently, the ultimate aim of an automized technique requires extensive, statistically confirmed series of experiments and long-term computerized memory banks of instantly recallable individual and group patterns. There are reasons to assume that such approaches would improve the reliability and make use of the great informative value of polyacrylamide gel electrophoresis for diagnosis and therapy control.

References

(1)   Maurer, H. R. and R. C. Allen, Z. klin. Chem. u. klin. Bichem. *10*, 220 (1972).
(2)   Maurer, H. R. and R. C. Allen, Clin. Chim. Acta *40*, 359 (1972).
(3)   Maurer, H. R., Disc Electrophoresis and Related Techniques of Polyacrylamide Gel Electrophoresis, Walter de Gruyter, Berlin–New York 1972, 2nd revised and expanded edition, p. 137.
(4)   Chrambach, A. and D. Rodbard, Science *172*, 440 (1971).

2.2 Quantitative Polyacrylamide Gel Electrophoresis: Mathematical and Statistical Analysis of Data

D. Rodbard and *A. Chrambach*

Abstract

The theoretical analysis of polyacrylamide gels in terms of the Ogston theory, and the application of moving boundary theory to buffer systems containing multiple weak acids or bases has made it possible to derive quantitative information from polyacrylamide gel electrophoresis (PAGE). This information is useful in describing both the physical properties of the macromolecule under study, the properties of the gel, and the optimal conditions of fractionation.

Once the conditions of PAGE (pH, temperature, polymerization reaction, current etc.) have been standardized and made reproducible, the only experimental measurement required to obtain this quantitative information is that of the relative electrophoretic mobility (R_f) at several known (or measured) gel concentrations (%T). These data are automatically processed by a series of computer programs, which provide 1. estimates of the reproducibility of the R_f's, 2. tests for identity/nonidentity of two proteins based on their R_f's, 3. estimates of the slope (K_R) and intercept (Y_0) of the Ferguson plot (log R_f vs %T), 4. joint 95% confidence limits for K_R and Y_0, 5. tests for identity/nonidentity of two proteins based on their Ferguson plots, 6. estimates of the molecular radius and apparent molecular weight, 7. free electrophoretic mobility and molecular net charge, 8. predictions of the gel concentrations providing optimal resolution from adjacent contaminants, 9. computer simulation for optimizing conditions of preparative scale PAGE.

The assumptions, limitations, problems, performance and potentialities of the first 7 of these methods will be discussed here. In particular, this report will review and expand on previous treatments regarding the testing of identity and homogeneity of two proteins on the basis of R_f's and Ferguson plots. Optimization of gel concentration and gel length with regard to separation and resolution in analytical PAGE, and to resolving rate in preparative PAGE (items 8 and 9 above) are the subject of the next chapter.

Emphasis is placed on recent developments and on the remaining assumptions, problems, and limitations of this approach to PAGE.

Introduction

The term "quantitative PAGE" implies that an attempt has been made to conduct fractionation effectively on a rational, scientific basis, to choose methods and optimum conditions on the basis of theoretical prediction, and to evaluate results quantitatively with automated data processing and statistical analysis. This concept was applied to what at the present time appears to be the fractionation method of choice — polyacrylamide gel electrophoresis (PAGE). More than any other presently available separation tool, PAGE excels in universal applicability to all charged molecules, in economy of sample and in resolving power (1). It is a preparative as well as an analytical method, although the development and application of the preparative scale apparatus and methodology has severely lagged behind the general acceptance of analytical PAGE.

We will attempt to show that analytical PAGE can be standardized and automated to give reproducible information concerning the physical properties of the molecule under investigation and to predict optimal conditions for fractionation. The prediction of optimal conditions for preparative PAGE (gel volume, elution buffer flow rate etc.) has been initiated but remains rudimentary until more information concerning bandwidth in polyacrylamide gels is available i. e. until the relation between apparent diffusion (dispersion) coefficient and gel concentration is further elucidated (2). The development of continuous scanning techniques for PAGE (3,53) promises to provide this information in the near future. One can therefore look forward to a standardized and automated methodology of preparative PAGE.

"Quantitative" PAGE implies that fractionation conditions be precisely known, both with respect to the nature of the gel and with respect to the buffer system.

Pore: The study of the *pore* of polyacrylamide gels is still in very early stages, both theoretically and experimentally. At the present time, in the absence of a simpler model, we still operate with the working hypothesis that the gel is composed of a random meshwork of linear fibers. Then, the distribution of *pore sizes* is given by the Ogston model (4, 5). But even this obviously primitive, limited model yields useful corollaries that can be verified empirically (6).

Multiphasic Buffer Systems: An exact and exhaustive theory of multiphasic buffer systems (MBS) (also designated as multiphasic zone electrophoresis (MZE) (7)), applicable to PAGE, was first formulated by Jovin 8 years ago, but published only recently (7, 8). A computer program (9) based on this theory provides a detailed description of discontinuous buffer systems operative at any pH (10).

An important consequence of the application of gel theory and buffer theory to the practice of PAGE lies in the considerable flexibility obtained. The original (11, 12) and still predominant practice of PAGE employs a single pore size, a single pH, and a single, uncontrolled temperature. In contrast, quantitative PAGE derives most of its resolving power from the use of variable and mathematically defined "optimal pore sizes" and from its use at pH values which provide fractionation on the basis of molecular net charge in addition to that provided by a restrictive gel pore.

We will review the information concerning the properties of macromolecules which can be derived from quantitative PAGE, as well as the application of such physical-chemical information to the optimization of analytical and preparative PAGE conditions. The relevant parameters are:

Relative Mobility (R_f)
Retardation Coefficient (K_R)
Relative Free Mobility (Y_0)

Molecular Radius ($\overline{R}$)
Molecular Weight (MW)
Free Mobility (M_0)
Net Charge (V)
Apparent Diffusion Coefficient (D)
Number of Theoretical Plates (N)
Identity between 2 species of macromolecules
Optimal gel concentration for analytical PAGE (T_{opt})
Optimal gel concentration and gel length for preparative PAGE

Some unresolved problems, limitations, and assumptions will be discussed in each category.

Mobility: M and R_f

Measurement: The first, and most important step in the use of polyacrylamide gel electrophoresis as a quantitative physical-chemical tool, is the measurement of either relative or absolute mobility, R_f or M (in contrast to the customary practice of qualitative, subjective inspection of the band pattern for analysis and interpretation of results). By measurement of R_f, one can directly ascertain the reproducibility of the fractionation system. This in turn, leads to the development of methodology to improve reproducibility and constancy of conditions (e. g. rigorous control of temperature, pH, conductance (13), polymerization efficiency (14) etc.).

The mobility of the species of interest (which we shall call the *Protein,* even though it may be a nucleic acid, mucopolysaccharide, or any other charged

species) may be measured in absolute terms, i. e. $cm^2 sec^{-1} volt^{-1}$, as by
Morris (15). Here, it is customary to calculate absolute mobility as

$$M = \frac{x \, l_1}{E \, t \, l_2} \tag{1}$$

where

M = mobility ($cm^2 sec^{-1} volt^{-1}$)
x = position of protein band (stained gel) (cm)
l_1 = length of gel, unstained
l_2 = length of gel, stained
E = voltage gradient in the gel, volts/cm
t = duration of electrophoresis (sec).

There are two methods for determining E. One uses direct measurement of
voltage drop across the gel by means of electrodes at the ends of the gel, and
the use of a very high impedence voltmeter (15). The other is based on
measurement of current (i), cross-sectional area of the gel (c), and conductance (κ), since

$$E = \frac{i/c}{\kappa}$$

$$volts/cm = \frac{amps/cm^2}{mhos/cm}$$

The latter approach is subject to systematic errors in the measurement of κ,
since this is affected by the gel, the residue of the ionic polymerization
catalysts, etc. Both approaches to measurement of E are subject to errors,
since E, i and κ usually vary during the course of the electrophoretic experiment. Even when the absolute mobility is measured directly (15), there are
several sources of error which may exceed those encountered in the measurement of relative migration distances along the gel. Further, Morris' method
does not seem readily applicable to multiphasic buffer systems. Accordingly,
it is restricted to samples which are sufficiently concentrated so that they can
be applied in a very small volume, within a relatively thin starting zone.
(Nevertheless, the starting zone will be wider than the width of the starting
zone in a multiphasic buffer system (7,11).)

Usually, the measurement of *relative mobility*, R_f, is more convenient than
measurement of absolute mobility. The mobility may be measured with
reference to one or two protein markers, as done by Ferguson (16) and
recently by Peacock (17). More commonly, small charged molecules, e. g. the
tracking dyes such as bromphenolblue and methylgreen may be used as
markers of moving boundaries in MBS (fig. 1). Fluorescent indicators (T. M.
Jovin, pers. comm.) of the naphthyl amine type or pyronin Y-dodecylsulf-

D. Rodbard and A. Chrambach

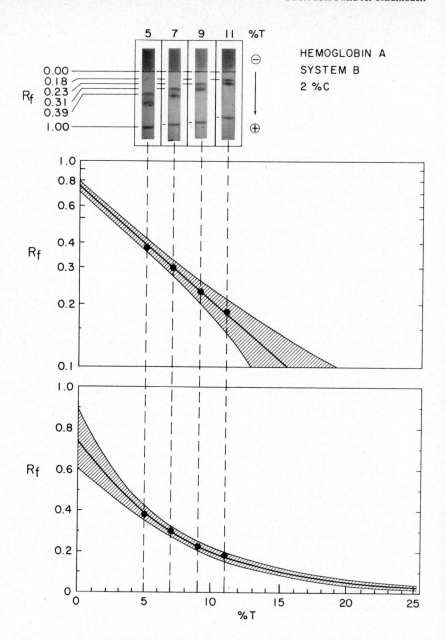

Fig. 1. R_f values of hemoglobin A in PAGE (system B, 5–11 %T, 2 %C) (Top). Plot of log R_f vs %T (Ferguson plot) of the same data (center). Plot of R_f vs %T of the same data (Bottom). R_f values for hemoglobin C (top, center band) are not calculated or plotted.

onate (V. Neuhoff, pers. comm.) have also been used for this purpose. Here, the relative mobility is defined by

$$R_f = \frac{\text{distance migrated by protein}}{\text{distance migrated by front}} = \frac{x}{f} \qquad (2)$$

However, since "markers" cannot usually be fixed, the position of the front, in practice, is measured on unstained gels, whereas the position of the protein is usually measured on stained gels. The R_f is then calculated as

$$R_f = \frac{x_2}{l_2} \frac{l_1}{x_1} \qquad (3)$$

$$= \frac{x_2}{f}, \quad \text{where } f = x_1 \frac{l_2}{l_1}$$

where

$x_1 =$ position of front, unstained gel
$x_2 =$ position of protein, stained gel
$l_1 =$ length of gel, unstained
$l_2 =$ length of gel, stained.

In a continuous buffer phase, the migration velocity of the dye is governed by its intrinsic mobility and depends on the gel concentration. In contrast, in multiphasic buffer systems, the dye is selected so that it is stacked for a wide range of gel concentrations (e. g. up to 15%T): accordingly, the velocity of the dye is determined by the properties of the *discontinuous* buffer system used. However, at very high gel concentrations (depending on the pH of the buffer system and the pK's of the dye), the dye may become *unstacked* due to molecular sieving. When this occurs, then the dye is no longer a satisfactory marker of the moving boundary constituting the *front,* and an additional correction factor must be introduced (see Appendix I):

$$R_f = \frac{x_2}{l_2} \cdot \frac{l_1}{x_1} \cdot R_f\{dye\} \qquad (4)$$

Methods for direct measurement of the position of the true front in multiphasic buffer systems have been described previously (e. g. slicing of the gel and measurement of the pH, conductance, or constituent ion concentration profile along the gel, immersion of the gel into a reagent which reacts with and demonstrates one of the buffer constitutents − e. g. $AgNO_3$ for Cl^-, La^{+++} for $HPO_4^=$, benzidine for persulfate etc.) (13). By such methods the $R_f\{dye\}$ relative to the moving boundary in front of it may be measured at several %T values. The Ferguson plot for the dye is constructed, and the R_f value predicted by the regression line is used as the correction factor in equ. 4.

Reproducibility, Precision: An important step in the "quality control" of a gel electrophoresis (PAGE) system, is to ascertain reproducibility of the R_f measurement. It is necessary to establish the standard deviation (σ) or the square of the standard deviation (σ^2) of R_f for replicate gels. The magnitude of this error will depend on whether the gels are run in the same experiment (same batches of reagents, identical polymerization conditions, etc.), or whether the gels are run in different experiments, using different reagents and conditions (e. g. minor variation in temperature, pH, ionic strength, catalyst concentrations, extent of de-gassing of polymerization mixture, or other factors leading to variable efficiency of the polymerization reaction). Thus, it is necessary to evaluate the precision of R_f within-experiments and the precision of R_f between-experiments. Furthermore, the magnitude of the random error in R_f depends on the method used for detection of the *Protein* band. Generally speaking, the magnitude of the random error in R_f is minimized when using staining of the gel and photography or densitometry; an R_f of 0.60 ± 0.01 might be representative (18, 19). When using slicing of the gel, combined with assay for radioactivity, immunological, biological or enzymatic activity, the magnitude of the standard deviation of R_f is usually increased by at least a factor of two, i. e. a standard deviation of ± 0.02 would be representative (20). Further, the magnitude of the standard deviation depends on the position on the R_f scale (from zero to one) (18). For example, an R_f of 0.1 might have a standard error of ± 0.008, while an R_f of 0.9 might have a standard error of 0.012. Thus, in general, the higher the R_f, the greater the absolute value of its standard deviation, but the smaller its relative error (coefficient of variation), and the smaller the error in the logarithm of R_f.

When all relevant comparisons are to be made on proteins run in the same gel (e. g. a gel slab, or when several proteins have been applied to the same gel tube), then the relevant parameter is the standard error of the *difference* in R_f (or alternatively, the standard error of the ratio of the mobilities of the two species). The standard errors of the difference (or ratio) of the R_f's from the same gel are usually considerably smaller than when the two R_f's are measured in different gels or in different experiments (19). For example, if the pH or gel concentration is slightly in error, or if there is an error in the determination of the position of the *front* or the length of the gels, then it is likely that the R_f's of two proteins in the same gel will be affected in the same direction, if not exactly to the same extent.

The measurement of the *precision,* i. e. the standard deviations of R_f within gels, within experiments, between experiments, as a function of R_f and the method of detection of the protein, need not involve an undue amount of experimental labor. By running a series of proteins (with R_f's over a fairly wide range) in duplicate in a series of experiments, it is possible to obtain the necessary information. The number of replicates which must be run can be considerably reduced by use of analysis of variance (ANOVA), i. e. pooling of information from various proteins and from various experiments.

When a sample R_f has been determined in duplicate, one can calculate the square of the standard deviation, with 1 degree of freedom (df):

$$s^2_{R_f} = \frac{(Y_1 - Y_2)^2}{2} = \frac{(\Delta R_f)^2}{2} \tag{5}$$

where Y_1, Y_2 represent the two measurements. These values for $s^2_{R_f}$ may be averaged, pooling data from several *Proteins* and experiments, for any narrow region of R_f:

$$\bar{s}^2 = \frac{\Sigma s^2 (n-1)}{\Sigma(n-1)} \tag{6}$$

One can plot the square of the standard deviation of R_f versus (R_f) for a series of proteins. Usually, one will observe considerable scatter, due to the random sampling error in the sample variances based on a small number of degrees of freedom. However, one can nevertheless draw a smooth curve, e. g. a parabola, through the scattergram (fig. 2). The value predicted by this smooth function should be a more reliable estimate of the scatter or uncertainty in the R_f, than the *local* or *point* estimate based on the replicates with R_f's at only one particular spot in the R_f spectrum.

If measurements are based on equ. 3, then we can predict the form of this relationship:

$$\sigma^2_{R_f} \cong \sigma^2_0 (1 + 3 R_f^2) \tag{7}$$

where $l_1 \sim l_2 \sim x_1 \sim 1$.

If measurements are based on equ. 2, then

$$\sigma^2_{R_f} \cong \sigma^2_0 (1 + R_f^2) \tag{8}$$

where $f \sim 1$.

If measurements are based on equ. 1, then

$$\sigma^2_M \cong \sigma^2_0 M^2 \left(\frac{1}{x^2} + \frac{1}{l_1^2} + \frac{1}{l_2^2} \right) \tag{9}$$

when E, t are known perfectly.

In general, as an approximation which is valid when the coefficient of variation of R_f or M is small,

$$\text{Var} (\log_e (R_f)) \cong \text{Var}(R_f)/R_f^2$$
or,
$$\text{Var} (\log_e M) \cong \text{Var}(M)/M^2 \tag{10}$$

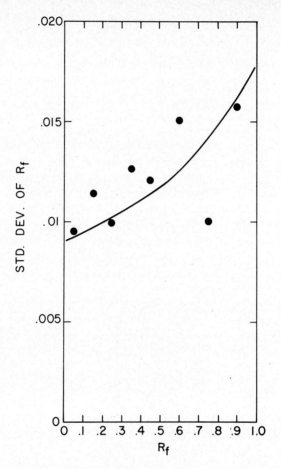

Fig. 2. Standard deviation of R_f (σ_{R_f}) vs R_f of hGH (system D, 10 % T, 2 % C) (18).

By suitable pooling of information (over experiments, over proteins, over R_f values), one can obtain reliable estimates of the precision of the R_f measurement with a minimum of experimental labor beyond that necessary to obtain the desired relative mobilities. The inclusion of one or two *standard* proteins in every experiment (e. g. one or two of the twelve gels in an electrophoresis tank, or one or two slots in a gel slab) can serve a dual purpose: To provide a check on systematic errors in experimental conditions, and to provide the data necessary for calculation of both within – and between – experimental error.

Problems with regard to R_f measurement at specific gel concentrations remain. The R_f has generally been measured manually with satisfactory accuracy and precision by use of caliper, ruler and magnifying glass (13). It is desirable to make these measurements on photographs of gels, rather than on the original gel. This provides a permanent record, permits repeated measurements and is usually necessary to obtain correction factors for gel shrinking/swelling during fixation and staining. A recording potentiometer with a mechanical input and digital output, as used in Dr. T. M. Jovins's laboratory, provides a semi-automated approach to measuring R_f with improved accuracy and precision. Densitometry can also be used to determine R_f values accurately. Both densitometry and electronic measurement of migration distances permit direct interfacing with computer.

Testing of Protein identity on the basis of R_f.

The information on the precision of the R_f is necessary for testing the identity of two *Proteins*. If both proteins have been studied in the same experiment (or same series of experiments), then one can use a paired *t*-test to evaluate whether the difference in R_f's, ΔR_f, is different from zero, using the within-experiment error in ΔR_f as the basis for calculation of the denominator:

$$ t = \frac{\overline{\Delta R_f}}{\sigma_{\overline{\Delta R_f}}} \tag{11} $$

If the proteins have been studied in separate experiments, then the larger between-experiment error must be used, and the unpaired *t*-test must be used:

$$ t = \frac{(\overline{R_f})_1 - (\overline{R_f})_2}{s \left(\frac{1}{n_1} + \frac{1}{n_2}\right)^{1/2}} \tag{12} $$

where n_1, n_2 are the number of independent measurements of $(R_f)_1$ and $(R_f)_2$, respectively. When multiple comparisons are made, it may be necessary to use Dunnett's multiple *t*-test, to adjust confidence levels, or use a formal analysis of variance (ANOVA). To obtain the most sensitive test of identity between two *Protein* species, one can measure the entire concentration profile on the same gel, and then calculate the positions of the centroid of each peak and their confidence limits, and then test for identity by methods analogous to the *t*-test. Unfortunately, this often requires double-label experiments with radioisotopes, and entails the loss of information and resolution attendant to slicing of the gel.

Ferguson Plots: K_R *and* Y_0

We refer to the linear relation between log R_f and gel concentration, %T (figure 1), as the Ferguson plot, after K. A. Ferguson who introduced this relation in 1964 (16):

$$\log(R_f) = \log(Y_0) - K_R T$$
or
$$\log(M) = \log(M_0) - K_R T \tag{13}$$

where

    K_R = retardation coefficient
    Y_0 = extrapolated R_f when T = 0
    M_0 = extrapolated free mobility when T = 0
    (logarithms to base 10 used unless otherwise stated).

Numerous other relationships between either relative or absolute mobility and gel concentration have been proposed by other workers. For example,

    R_f vs. $1/T$

    R_f vs. $1/T^2$

    R_f vs. $1/\sqrt{T}$

    R_f vs. $\log(T)$.

Some of these other relationships have been useful, empirically, in providing linearization over a limited region of mobility or gel concentration. However, none of these other relationships have proven to be as effective or universally applicable as the one proposed by Ferguson (5, 13, 16). Further, these other relationships or models lead to non-linearities or to absurd predictions as % T approaches zero. Also, the Ferguson plot is compatible with, and may be regarded as a corollary of, Ogston's model for the calculation of the distribution of pore sizes in a random meshwork of fibers (5, 13). The Ferguson relationship is analogous to the linear relationship between $\log(K_{av})$ and %T for gel filtration (5, 22, 23), and to the linear relationship between log(D) and %T (2, 5). This relationship has been shown to be valid and applicable for molecules ranging in molecular weight from 300 to 2.2×10^6 (or higher), for proteins, both single and double stranded nucleic acids, acid mucopoly-saccharides, peptides, organic dyes and SDS-denatured proteins. The behavior of the diffusion coefficients of water, NaCl, and urea suggests that this relationship should also apply to molecules in that molecular weight range (24). Accordingly, in those few rare, instances where non-linearity of the Ferguson plot has been reported (e. g. (25)), one must carefully evaluate the data for possible sources of artifact, e. g. failure to refer R_f values to the identical moving boundary at low and at high %T (see above); failure to correct for unstacking of the tracking dye, errors in % T due to incomplete polymerization (see below), failure to stack the protein during the *stacking stage* of routine PAGE in MBS etc. Nevertheless, it is possible that the Ferguson relationship will fail: the failure of the relationship between $\log(K_{av})$ vs. T to extrapolate through the origin (i. e. $K_{av} = 1$ when $T = 0$) (5, 22) and the systematic relationship between Y_0 and %C (26) suggest that this may be the case.

The linear relationship between $\log(R_f)$ and gel concentration (%T) enables us to combine information from several gel concentrations, and to characterize each protein in terms of two parameters: K_R which is a measure of molecular size, and Y_0 which is (or can be translated into) a measure of molecular free mobility. These two parameters can be obtained from a graph of $\log(R_f)$ vs. %T, as has been used by most workers (13).

The precision of %T:

The precision of R_f has been discussed above. The uncertainty with regard to %T may be even more severe than the uncertainty with regard to R_f. Nearly everyone makes the implicit assumption that the polymerization reaction goes to completion. But it does not. Each of the three generally used catalysts — TEMED, K-persulfate (KP) and riboflavin (RN) — contribute to polymerization efficiency, at least at acid pH (14) and presumably at neutral pH. (At acid pH, KP is a relatively inefficient initiator (14); at alkaline pH, RN is relatively ineffectual). Depending on the relative concentration of each of the catalysts, polymerization efficiency may range anywhere from 50 to 99%, despite the fact that a *gel* has formed (figure 10 of (14)). This necessitates use of considerable caution in the interpretation of practically all reported R_f values and Ferguson plots in the literature. As a minimum, it is necessary to measure polymerization efficiency (%PE) for at least one, and preferably several gel concentrations in every new buffer system, or whenever the catalysts or their concentrations are changed. A general guide to polymerization has been reported (14).

Ferguson plot: Linear Regression Analysis

An objective rule for curve-fitting should be used to avoid personal biases; the least-squares or maximum-likelihood criteria seem to be most appropriate. The customary use of simple, *unweighted* least — squares linear regression between $\log R_f$ and %T can be seriously misleading, due to the severe non-uniformity of variance of $\log R_f$. As R_f becomes small (R_f approaches zero), the relative error or percentage error in R_f becomes enormous (even infinite), and so does the error in $\log R_f$ (cf. figure 2 and Equ. 10). Accordingly, it is necessary to use a *weighted* regression. Since the variance of log R_f is calculated as a function of R_f (not directly as a function of %T) (equ. 7–10), it is necessary to use an iterative approach, in order to re-adjust the weights after each successive approximation. Usually 3 or 4 iterations are sufficient: the method converges very rapidly. The results from an unweighted regression can be used as the initial estimates for the iterative curve-fitting procedure. Using the maximum-likelihood method of Finney (27), the regression analysis takes the following form:

$$y = a + b(x - \bar{x}) + \epsilon \tag{14}$$

$$y = \text{"working log"} = \log\hat{Y} + \frac{Y - \hat{Y}}{\hat{Y}} \frac{1}{\ln(10)} \tag{15}$$

where,

Y = observed R_f

$\hat{Y}$ = R_f predicted on the basis of the previous regression line

x = %T (observed)

$$a = \Sigma wy/\Sigma w = \bar{y} \tag{16}$$

$$b = \frac{\Sigma w(x - \bar{x})(y - \bar{y})}{\Sigma w(x - \bar{x})^2} \tag{17}$$

$$= -K_R$$

$$w = 1/\text{Var}(\log(R_f)), \text{ evaluated for } \hat{Y}, \tag{18}$$
from equ. 7–10,

e. g. using equations 7 and 10, one has

$$w = 1/(1 + 3/R_f^2) \tag{19}$$

and

$$\log Y_0 = a - b\bar{x} \tag{20}$$

$$Y_0 = \text{predicted (extrapolated) } R_f \text{ when } x = T = 0 \tag{21}$$
$$= 10^{\log Y_0}$$

In lieu of the linear relationship between log R_f and %T, one can obtain the parameters (K_R, Y_0) using general methods and computer programs for non-linear curve fitting, using the negative exponential relationship between R_f and %T (figure 1, bottom).

$$R_f = A\,e^{-\alpha T}, \text{ where } A = Y_0, \alpha = K_R \cdot \log_e 10 \tag{22}$$

However, once again, it is necessary to use weighting, since as mentioned above (e. g. equ. 7–9 and figure 2), the error in R_f is not constant, but varies slightly as a function of R_f (by perhaps a factor of 2 in terms of the standard deviation, or a factor of 4 in terms of the variance — and the weight assigned to each point). This approach appears to offer little or no real advantage, since the conceptual simplicity and familiarity of the linear regression methods are lost. In practice, we have preferred the use of the linearization by the log transformation, although the results of the two methods should be indistinguishable.

Pooling of Residual Variance (s_0^2) to Increase Reliability:

In practice, it is customary to construct Ferguson plots with 5 to 7 points on each curve. A minimum of three points is required for any form of statistical analysis of results. The number of degrees of freedom (df) is equal to the number of points on the curve (n) minus two: $df = n - 2$.

When one is dealing with a small number of degrees of freedom, the estimate of the residual variance (scatter around the curve), s_0^2, is very imprecise, since it is subject to a large sampling error. Accordingly, while the standard errors for K_R and Y_0 may remain small, the 95% confidence limits (CL) for these parameters will be very wide, since one will have a very large value for the appropriate t-or F-statistic:

$$s_{K_R}^2 = s_0^2 / \Sigma \, w(x - \overline{x})^2 \qquad (23)$$

$$s_{\log Y_0}^2 = s_0^2 \left(\frac{1}{\Sigma w} + \frac{\overline{x}^2}{\Sigma w(x - \overline{x})^2} \right) \qquad (24)$$

$$s_{Y_0} \cong s_{(\log Y_0)} \, Y_0 / \ln(10) \qquad (25)$$

$$(\log Y_0)_{u,l} = \widehat{(\log Y_0)} \pm t \, s_{\log Y_0} \qquad (26)$$

$$(Y_0)_{u,l} = 10^{(\log_{10} Y_0)_{u,l}} \qquad (27)$$

$$(K_R)_{u,l} = \hat{K}_R \pm t \, s_{K_R} \qquad (28)$$

Note: Apply CL symmetrically in terms of $\log Y_0$, then take antilogs.

$$\text{Width of confidence limits} \propto t \, s_0 / \sqrt{n} \qquad (29)$$

Further, the t-tests for identity of slope or intercept (considered separately), will also be insensitive, due to low df:

$$t = \frac{(K_R)_1 - (K_R)_2}{s_{\Delta K_R}} \qquad (30a)$$

$$s_{\Delta K_R} = s_0 \left(\Sigma \frac{\overline{x}^2}{\Sigma w(x - \overline{x})^2} \right)^{1/2} \qquad (30b)$$

$$t = \frac{(\log Y_0)_1 - (\log Y_0)_2}{s_{\Delta \log Y_0}} \tag{31a}$$

$$s_{\Delta \log Y_0} = s_0 \left(\Sigma \frac{1}{\Sigma w} + \Sigma \frac{\overline{x}_2^2}{\Sigma w (x - \overline{x})^2} \right)^{\frac{1}{2}} \tag{31b}$$

One way to overcome this difficulty is to pool information from several Ferguson plots. In a representative experiment, we might have 10 proteins, each characterized at each of 5 gel concentrations. Thus, the residual variance (s_0^2) for each Ferguson plot has only 3 df, and the appropriate values for t or F for calculation of 95% CL for K_R and Y_0 will be very large. However, we can examine the estimates of the residual variance for each of the proteins, and check whether these values appear *homogeneous*. A Bartlett's test for homogeneity of variance may be used, although this test is very inefficient and may be unreliable when one is dealing with a small number of degrees of freedom, and the probability levels for Bartlett's test may be severely perturbed if one has departures from the ideal Gaussian distribution. Accordingly, more sophisticated tests for homogeneity of variance may be used (51). When only two groups are being pooled (as for making t-tests for slope and intercept, when only 2 Ferguson plots are being considered), then the F-ratio may be used to test for homogeneity of variance, i. e.

$$F = s_1^2 / s_2^2 \tag{32}$$

Then, having decided that the estimates of residual variance are *compatible* or *homogeneous*, we can pool them, thus obtaining an estimate of the pooled residual variance with df = $\Sigma(n - 2)$:

$$\overline{s}_0^2 = \Sigma s^2 (n - 2) / \Sigma (n - 2) \tag{33}$$

Thus, pooling of information provides us with a situation with *large* rather than *very meager* df, exactly analogous to what was done when we pooled information regarding $s_{R_f}^2$ in equation 6. Accordingly, the confidence limits for K_R and Y_0 may be reduced substantially (e. g. often by a factor of two or more), and our ability to detect a significant difference in either or both of these parameters is markedly improved: We simply use $\overline{s}_0$ instead of s_0 in equations 23–31, and evaluate t values using the combined df.

The standard errors of K_R and Y_0 may then be recalculated from equ. 23–24 using the pooled residual variance, $\overline{s}_0^2$, in lieu of the s_0^2 from the individual Ferguson plots.

In pooling residual variances, as in the pooling of the standard deviation of R_f, it is necessary to avoid pooling information obtained by different methods (e. g. staining *vis a vis* slicing) whenever possible. When it is necessary to pool information from two proteins studied by two different methods (e. g. activity-assay and staining), it is usually preferable to pool data derived from each method separately. In the event that the residual variances are unequal, one may still be able to perform valid t-tests, using:

$$s^2_{\Delta K_R} = (s^2_{K_R})_1 + (s^2_{K_R})_2 \tag{34}$$

$$s^2_{\Delta Y_0} = (s^2_{\log Y_0})_1 + (s^2_{\log Y_0})_2 \tag{35}$$

in lieu of equ. 30b and 31b. Alternatively, the methods developed for comparison of two sets of data with unequal variance (e. g. based on the Fisher-Behrens distribution) may be used (27, 28).

Testing Identity of 2 Species on the Basis of Their Ferguson Plots:

In order to test the identity/non-identity of 2 protein species, we test whether or not the two corresponding Ferguson plots are *identical* or homogeneous. One can approach this question, by testing whether or not the K_R's or Y_0's are significantly different, e. g. by the conventional t-tests for slope and intercept (equ. 30 and 31) (28). The sensitivity of these tests is increased if we use the residual variance from several proteins (equ. 33), rather from just the 2 Ferguson plots being compared. However, the errors in K_R and Y_0 are highly correlated: If one has an erroneously high estimate of K_R (e. g. due to an error in one or a few of the points on the line), one is also likely to have an erroneously high estimate of the Y_0, and vice versa. Thus, when we are making a comparison between two proteins in terms of size (K_R) and in terms of free mobility (Y_0) by use of equ. 30 and 31, we are not using two independent criteria. Accordingly, each protein (or each Ferguson plot) can be characterized in terms of the joint 95% confidence limits for K_R and Y_0. These confidence limits are in the form of an ellipse, when plotted in terms of log Y_0 vs. K_R (figure 3, upper left). This ellipse is first defined in terms of an ellipse for $\bar{y} = a$ (equ. 16) and slope, which are subject to "independent" errors. This ellipse (figure 3, lower left) is defined by (28):

$$\frac{(a - \alpha)^2 \, \Sigma w + (b - \beta)^2 \, \Sigma w(x - \bar{x})^2}{2 \, \bar{s}^2_0} = F \tag{36}$$

an F ratio with 2 df in the numerator and with denominator df corresponding to $\bar{s}^2_0$, and a, b defined by equ. 16, 17; β = "true" slope, α = "true" expectation of y, given x's. Then, each point on this ellipse is *mapped* into a corresponding point on an ellipse in terms of the y-intercept (log Y_0) vs. slope (figure 3, upper left), using the relationship (28):

$$\eta_0 = \alpha - \beta \, \bar{x} \tag{37}$$

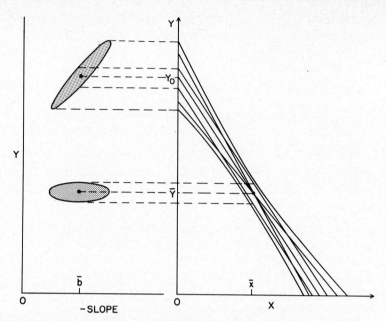

Fig. 3. Schematic representation of 95% confidence envelopes for K_R and Y_0 (left), in relation to the confidence limits for the regression line (right).

JOINT 95% CONFIDENCE ENVELOPES OF K_R AND Y_0
FOR hPR (■) AND hGH (▨) IN 3 PAGE SYSTEMS

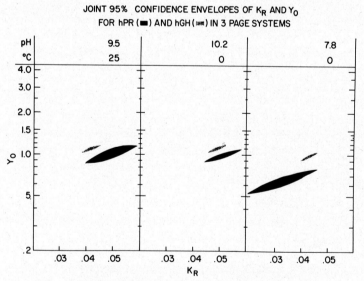

Fig. 4. Joint 95% confidence envelopes of K_R and Y_0 for hGH-B (shaded) and human prolactin (hPRL) (black). PAGE (system A) analysis was carried out for hGH with detection by staining, for hPR with detection by RIA (18, 31).

Note — this ellipse shows the possible combinations of slope and intercept allowed by the 95% CL for the line (figure 3), and is conceptually nearly equivalent to that confidence region. An example of this type of analysis is shown in figure 4.

We have arbitrarily used the 95% confidence limits for (K_R, Y_0), although any other probability level could arbitrarily be used, e. g. 50% confidence limits corresponding to the *probable error,* or the 88% confidence limit, etc. These joint CL for K_R and Y_0 are calculated at the same time, and by the same program used to calculate the Ferguson plot. These *original* K_R-Y_0 ellipses are usually very large, due to the small number of df corresponding to s_0^2 for a single Ferguson plot. If the pooled $\bar{s}_0^2$ is used, with its larger number of degrees of freedom, one obtains a *reduced* ellipse. These *reduced* ellipses are provided by a program (IDENT), which also tests the residual variance estimates for homogeneity (F test or Bartlett's test) and provides the estimate of the pooled residual variance (equ. 33).

Ellipse Overlap Criteria: When the 95% CL for two proteins do not overlap, then we can conclude that the proteins are significantly different with $P < (0.05)^2$ or 0.0025. More difficult to interpret are the cases of partial overlap. If two ellipses overlap, but neither one overlaps the center of the other one, we infer a statistically significant difference, $P < 0.05$. If either ellipse overlaps the center of the other ellipse, the result is not significant at the $P < 0.05$ level. If overlap of any point of the ellipse with the other is used as the criterion for identity, and one wishes to preserve the $P < 0.05$ level of significance, one should use the $F_{0.78}$ value for calculating the ellipses, rather than $F_{0.95}$. If we use the 78% CL for K_R, Y_0, then two proteins with non-overlapping (but just touching) ellipses would be significantly different with an approximate value of $P \cong (0.22)^2 \cong 0.05$.

An F-test for Identity of 2 Ferguson Plots:

The above *overlap* criteria are useful in clear-cut cases, e. g. no overlap or complete overlap. In borderline cases, the interpretation of the partially overlapping ellipses is hazardous. Accordingly, it is useful to have a simple, unequivocal numerical statistical test for identity of two Ferguson plots. This criterion for non-identity of two proteins (simultaneous consideration of K_R and Y_0) can be expressed in terms of an "F-test", performed as follows:

1. Calculate the regression lines for the two proteins which are being compared and estimate the residual variance for each. Test to see whether the residual variance is equal for both curves (equ. 32); if so, combine the two estimates, weighted according to their respective df (equ. 33), to obtain $\bar{s}_0^2$.

2. Now pool all of the data (i. e. all pairs of T, R_f values), and recalculate the regression line. If the two proteins were really the same, then this combined regression line should be the "correct" one, or at least a better estimate than either of the two separate lines, since it is based on more points. Again, recalculate the residual variance around the line. By the use of only one regression line, we have reduced the number of parameters from four to two (i. e. we now have only a single K_R and a single Y_0, instead of two of each). However, we have increased the scatter around the line (in terms of sums of squares of deviations around the line, SS). Only in the case where the K_R's and Y_0's for the two curves are exactly the same, would there be no increase in the residual SS. The question is, does the reduction in the number of parameters by 2 (with an increase in 2 df in the residual variance) result in a significant increase in the (weighted) residual sum of squares of deviations around the line? Equivalently, does the change from 2 parameters to 4 parameters result in a significant reduction in the scatter (residual variance) around the curve?[1]. The appropriate F ratio is:

$$F = \frac{\Delta SS/\Delta df}{\bar{s}_0^2} \tag{41}$$

where

$$\Delta SS = SS_3 - (SS_1 + SS_2)$$
$$\Delta df = 2$$
$$\bar{s}_0^2 = (SS_1 + SS_2)/(df_1 + df_2)$$
$$SS = \Sigma w(y - \bar{y})^2 - b^2 \Sigma w(x - \bar{x})^2$$

This question differs from the one usually posed in statistics texts which use a two step test procedure: 1. are the lines parallel? (or are the slopes identical?) based on the t-test for slopes (equ. 30), followed by 2. are the intercepts equal – subject to the hypothesis that the lines are parallel? The latter involves recalculation of the residual variance, under the hypothesis that the slopes are equal, i. e.

$$s_0^2 = \Sigma\Sigma w(y - \bar{y})^2 - \bar{b}^2 \Sigma\Sigma w(x - \bar{x})^2 \tag{38}$$

where

$$\bar{b} = \Sigma\Sigma w(x - \bar{x})(y - \bar{y})/\Sigma\Sigma w(x - \bar{x})^2 \tag{39}$$

Then, the t-test for identity is given by:

$$t = \frac{(\log Y_0)_1 - (\log Y_0)_2}{s_0 \left(\Sigma \frac{1}{\Sigma w} + (\bar{x}_1 - \bar{x}_2)^2/\Sigma\Sigma w(x - \bar{x})^2\right)^{1/2}} \tag{40}$$

This is because, when dealing with Ferguson plots, we have no *a priori* reason to assume that the lines are parallel, except in a few special cases (e. g. hemoglobins, desamidospecies, etc.).

with numerator df = 2, denominator df corresponding to $\bar{s}_0^2$. Subscripts 1, 2 refer to the two species considered separately, and the subscript 3 refers to the regression line calculated on the basis of all of the data.

A simplified Procedure for testing identity of 2 Ferguson plots:

The procedure described above utilized three separate regression analyses: for curve 1, curve 2 and for both curves simultaneously. In practice, it is necessary only to perform the first two regression analyses. The results for the combined regression analysis, and the test of identity, can then be calculated from intermediate results of these two regressions,

$$\bar{x}, \bar{y}, K_R, \Sigma w(x - \bar{x})^2, \Sigma w(y - \bar{y})^2, s_0^2, n, \Sigma w,$$

using formulas developed by J. Gart (pers. comm.).

$$\Delta a' = (\log Y_0)_1 - (\log Y_0)_2$$

$$\Delta b = (K_R)_1 - (K_R)_2$$

$$C_{11} = \Sigma \frac{1}{\Sigma w} + \Sigma \frac{\bar{x}^2}{\Sigma w(x - \bar{x})^2}$$

$$C_{22} = \Sigma \frac{1}{\Sigma w(x - \bar{x})^2}$$

$$C_{12} = \Sigma \frac{-\bar{x}}{\Sigma w(x - \bar{x})^2} \tag{42}$$

$$D = C_{11} C_{22} - C_{12}^2$$

$$C^{11} = C_{22}/D$$

$$C^{12} = -C_{12}/D$$

$$C^{22} = C_{11}/D \tag{43}$$

$$Q = C^{11} (\Delta a')^2 + C^{22} (\Delta b)^2 + 2C^{12} (\Delta a') (\Delta b)$$

$$F = \frac{Q/2}{\bar{s}_0^2} \tag{44}$$

df (numerator) = 2;
df (denominator) = same as for $\bar{s}_0^2$

These calculations are performed by program IDENT, in the following sequence:

1. The residual variances from the regression of log R_f vs. T from any number of proteins are analyzed for homogeneity by a Bartlett's test and compared with the appropriate chi-square statistic.

2. The user is then given a choice between *pooling* data from all proteins, or analyzing the data from each protein (or pair of proteins) separately.

3. Joint 95% CL ellipses for K_R and Y_0 are provided for each protein, utilizing either the individual or the pooled residual variance, as requested.

4. Tests of identity for any two species are then performed (equ. 42–44), either utilizing $\bar{s}_0^2$ from just the two *Proteins* being compared, or the pooled $\bar{s}_0^2$ for all *Proteins*. These F tests are interpreted as significant or not, by calculation of the corresponding P value.

This method has several advantages: it is objective, and based on standard, conventional statistical methods. It permits us to combine information from several proteins (to obtain an estimate of the pooled residual variance), and it permits us to analyze information from all gel concentrations studied (in contrast to performing a separate analysis such as a *t*-test on R_f's for each different gel concentration).

However, these methods for testing identity of proteins have certain drawbacks that must be borne in mind by the investigator:

To construct Ferguson plots for several proteins, it is usually necessary to perform several experiments, often over a fairly lengthy period of time (especially when attempting to obtain 7 or more points on each curve). Accordingly, the *between-experiment* error is introduced, and this may be several times larger than the within-experiment error.

Second, the present statistical analysis makes no distinction between the error in R_f's obtained at the same gel concentration (replicate gels) and the scatter around the line of the means of the replicates for each gel concentration. To be sure, the regression analysis could be modified to include this type of analysis of variance. However, this would greatly increase the complexity of interpretation of results, and decrease the number of degrees of freedom, if variation of R_f's around the line between gel concentrations is significantly greater than variation amoung R_f's for any given %T, as expected if %PE (polymerization efficiency) depends on %T.

Third, the present analysis is able to consider only random errors (although departures from a Gaussian distribution are not too serious). However, these analyses provide no safeguard against the presence of systematic errors. For example, if the pH is in error during the analysis of one of the two proteins, then its Y_0 may be markedly affected, and it might appear significantly different from another protein studied in a different experiment. Of course, inclusion of a series of *standard* proteins and analysis of its R_f's, K_R's, Y_0's might provide a warning in this case. Also, this points out the desirability of performing electrophoresis on the two proteins under *identical* conditions in the same experiment, or using methods of experimental design such as *random blocks,* to allocate proteins and gel concentrations to experiments.

Accordingly, the *identity test* described above must be regarded as provisory. In general, it will be desirable to confirm the results of such a test, by means of an experiment wherein all necessary comparisons are made within the same *run,* using the optimal gel concentration for that comparison. A general strategy may be formulated as follows:

1. Construct a Ferguson plot for each of the proteins of interest.

2. Test *identity* on the basis of K_R and Y_0 separately (equ. 30, 31) and on the basis of the K_R, Y_0 ellipses (equ. 36–37) or the F-test for K_R-Y_0 considered simultaneously (equ. 42–44).

3. Compare R_f's at single gel concentrations within the same experiment (equ. 11, 12).

4. Calculate the gel concentrations for maximal separation (T_{max}) and optimal resolution (T_{opt}) (5, 20, 29, 30), using program T-OPT (30).

5. Compare R_f's obtained at this T_{opt} within the same experiment (equ. 11, 12).

6. Repeat these analyses under at least two different sets of conditions differing as widely as possible in pH, buffer composition, temperature, etc.

Testing Homogeneity of a Single Preparation:

Closely related to the problem of testing identity of two different species, is the problem of testing homogeneity ("purity") of a given protein preparation on the basis of R_f, K_R and Y_0. It should be realized, however, that electrophoretic homogeneity may be an excessively stringent requirement for many purposes e. g. sequence analyses of proteins. A number of reactions in protein chemistry, e. g. 1. deamidation, 2. aggregation, 3. desialylation of glycoproteins, 4. binding of fatty acids or other ligands, 5. presence of several conformational states, or 6. cleavage of peptide bonds without release of fragments (31) cause electrophoretic heterogeneity without seriously interfering with particular studies such as protein sequencing or derivatizations.

Obviously, a finding of 2 bands or more on a single gel excludes homogeneity. However, contrary to popular belief and practice, the finding of a *single* band under a single set of buffer and gel conditions, does *not* necessarily establish homogeneity. The pH or the gel concentration or the ionic strength etc. may simply be insufficiently optimized for effective resolution. We have suggested an operational rule that PAGE analysis at 3 divergent pH's and over a wide range of gel concentrations, with a single band (R_f) at each %T, and 3 linear Ferguson plots, provides a sufficient criterion of homogeneity (1).

In addition to the variation of %T, pH, temperature, etc. and analysis of the Ferguson plot as described above, it is desirable to examine the entire concentration profile of the peak. Is there significant departure from a Gaussian distribution? Is there evidence for a shoulder (e. g. on a densitometry pattern or in profiles of radioactivity obtained from slices)? What is the standard deviation of band width? This value should be calculated as the square root of the second moment around the mean. Alternatively, for a Gaussian peak, one can use the relationships:

$$\sigma^2 = \frac{w^2}{5.55} \tag{45}$$

where w is the width of the peak at one-half maximal height, or

$$\sigma = \frac{A}{\sqrt{2\pi}\,h} \tag{46}$$

where

A = area of the peak (above baseline)
h = maximal peak height

All of these relationships require accurate estimation of the baseline. Then, the number of theoretical plates (5, 32) may bé calculated as

$$N = \frac{x^2}{\sigma^2} \tag{47}$$

where x is the distance migrated (x and σ must be in the same units).

The use of "N" makes the implicit assumption that $\sigma \propto M$, which is not exactly the case. Accordingly, "N" may vary with the R_f. Thus, it seems preferable to examine the empirical relationship between σ^2 and R_f, for a series of proteins of known (or presumed) homogeneity. Then, the band width of the unknown (unknown with respect to homogeneity) can be compared with the series of standards. The *standards* should have the same general physical properties, in terms of diffusion coefficients and particularly net charge, as the unknowns. If the *unknown* shows markedly greater σ^2 than the appropriate *standards*, this is suggestive evidence for heterogeneity of the protein. Conversely, if the *unknown* is in line with the reference species, then we have presumptive evidence for homogeneity. Many factors affect bandwidth (e. g. concentration dependence of the diffusion coefficient), and the above rules may be misleading at times. Nevertheless, this type of analysis is simple, can be obtained without performing additional experiments, does NOT require a computer, and can be highly informative. Thus, although the information in the bandwidth (and skewing) has generally been neglected, it may be of critical importance when evaluating homogeneity.

Note that the tests for identity and homogeneity should be performed in terms of R_f, Y_0, and K_R, and not in terms of secondarily calculated variables such as radius, molecular weight, free mobility, net charge, etc.

Molecular Radius and Weight: $K_R \to \bar{R} \to MW$

The K_R is a measure of molecular size. However, for reasons of inter-convertibility of data with other methods, it is frequently desirable to trans-late this parameter into terms of molecular radius and molecular weight. For globular proteins, this presents no great difficulty: there is a linear relation-ship between the square root of the retardation coefficient and the cube root of molecular weight, i. e. the molecular radius (figure 5a) (5, 13, 15, 16).

$$\bar{R} = a + b \sqrt{K_R} \qquad (48)$$

The $\bar{R}$ values for the *standard* proteins on the calibration curve are calculated as:

$$\bar{R} = \left(\frac{3 \, MW \, \bar{v}}{4 \, \pi \, N} \right)^{1/3} \qquad (49)$$

where

N = Avogadro's number
$\bar{v}$ = partial specific volume.

After analysis of literally hundreds of calibration curves of this type, obtained under a wide variety of conditions of pH, temperature, ionic strength, and buffer constituents, it appears that this method is superior to the use of a linear relationship between molecular weight (MW) and retarda-tion coefficient as originally advocated by Ferguson (16), Hedrick and Smith (33) and others (figure 5b). The use of a general nonlinear function to relate K_R and the square of molecular radius (R^2 or $MW^{2/3}$) can also be used, but sacrifices the simplicity and familiarity of straight-forward linear regression:

$$K_R = a + b \, MW^{1/3} + c MW^{2/3} \qquad (50)$$

The type of calibration curve given by (equ. 48) and figure 5a has been shown to be linear, and "valid" for a MW range from 300 to 2×10^6. Further, it appears that unweighted regression can be used, since the cube-root transfor-mation of molecular weight very nearly compensates for the fact that high MW species usually have a larger standard deviation, but smaller coefficient of variation than low MW species.

This *Ferguson plot* method for determination of $\bar{R}$ and MW by PAGE is especially useful in situations where other methods are not applicable: viz: when one is dealing with microgram amounts in heterogeneous systems, or when one is interested in molecular size as a function of ionic milieu. The

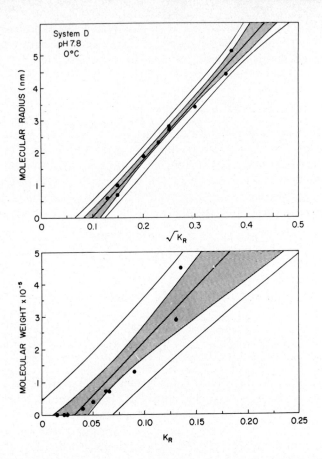

Fig. 5. Standard curves for system D of $\bar{R}$ vs $\sqrt{K_R}$ (figure 5a, top) and MW vs K_R (figure 5b, bottom). Standard proteins and their MW's (in parentheses) are: Bromphenol-blue (670), Bacitracin (1411), Glucagon (3485), Soybean Trypsininhibitor (22700), Ovalbumin (43500), BSA monomer (67000), Transferrin (74000), BSA dimer (134000), Phycoerythrin (290000), Ferritin (450000). Although use of MW vs. K_R is justified for random coiled, e. g. SDS denatured proteins, the use of $\bar{R}$ vs. $\sqrt{K_R}$ is distinctly preferable for globular proteins, as seen here. Figure 5b shows non-linearity and much wider confidence limits than figure 5a.

magnitude of the error in the calibration curve can be reduced by the simple expedient of increasing the number of standard proteins studied. Separate calibration curves must be established for every new set of experimental conditions (pH, temperature, ionic strength, % crosslinking, etc.).

Our programs for estimation of molecular radius (13) also provide an estimate of the standard error of $\bar{R}$, and the 95% CL for $\bar{R}$ (28). These values may then be converted into the corresponding molecular weight, with its standard error and 95% CL. Usually, these standard errors appear enormous compared with those reported by other methods (18, 20, 34–41). However, most of this discrepancy can be explained on the basis of the type of statistical analyses selected, rather than on the basis of the performance of the method in general. For instance, our standard errors in $\overline{R}$ and MW include the error or uncertainty in the location (slope, intercept) of the calibration curve. In contrast, other workers have drawn their calibration curves *by eye*, and then (implicitly) assumed that the line is known perfectly (even though it may consist of 4 or 5 points with only 2 or 3 df). Then, they have examined the *error* in the K_R, and calculated the corresponding change in $\overline{R}$ or MW, while ignoring errors in the estimation of the regression line. Further, many workers have used ± 1 standard deviation or the probable error, ± 0.6 std. deviations, whereas we use ± 2 standard deviations when we have a *large* (greater than 20) number of degrees of freedom, and ± $t_{.975}$ standard deviations in general.

Thus, the large errors of our method may be no worse than those of other methods (e. g. we may be looking at the difference between + and – 3 standard errors, i. e. 6 standard errors, and may be comparing it with 0.6 standard errors).

Also, in our studies, we have not yielded to the temptation of rejecting *outliers*. Often, only one or two of several (e. g. 10) standard proteins will contribute as much as 50% of the total scatter around the line (calibration curve). The customary practice, both in PAGE, SDS-PAGE, or gel filtration is to reject these points. This can have a dramatic salutary effect on the size of the calculated confidence limits, but does not improve the reliability of the results. Alternatively, many investigators then correct the MW of the anomalous protein (e. g. by reasoning: "It may be the dimer", "It may be the subunit", etc.). Since one has no assurance that similar effects are not also present for the unknown, no data points should be rejected, unless one has independent data to justify this (e. g. from sedimentation analysis).

Nevertheless, it is true that errors in the estimates of MW from PAGE are large. This appears to be due to variation in hydration, partial specific volume, and the shape (conformation) of various proteins (13). Methods using denaturing detergents (e. g. sodium dodecyl- or decyl-sulfonate — designated SDS and NaDS respectively) should provide a much better correlation with MW, provided that conditions are chosen such that these detergents confer equal charge density or free mobility to each protein under study (21, 25, 42). However, even when one is dealing with SDS denatured proteins, it is still imperative to construct a Ferguson plot at least once in order to check that the Y_0 is constant for all species.

Non-spherical Macromolecules: For non-spherical molecules (e. g. nucleic acids), estimation of molecular weight and radius is far more complex (e. g. see (13, 21, 25)). It is still not established, how to estimate the effective radius for a non-spherical molecule. Many workers have assumed that the "Stokes" radius R_s should be used. However, this is by no means established, and there is considerable evidence that this might not be the case (13). Based on the "Ogston" model for a gel, we might expect that the effective radius of a molecule depends on its surface area (when one is dealing with a gel composed of a random suspension of linear *fibers*), or only on its volume (when one is dealing with an ideal gel composed of a random suspension of *points*) (5). However, experiments to attack this question have not been definitive (26). It appears that the finite effective radius of the polyacrylamide gel fiber may systematically change the relationship between K_R and molecular size and shape.

Analogy with Gel Filtration:

Many users of gel filtration have also assumed that K_{av} is a measure of Stokes' radius, and have converted K_{av}'s into apparent R_s values. Some workers have then converted these R_s values into diffusion coefficients (43). Next, these apparent diffusion coefficients have been combined with sedimentation coefficients in the Svedberg equation to calculate apparent molecular weight, as in the sedimentation velocity-diffusion method. Although Monty and Siegel (44) have shown that there is a *better* correlation between K_{av} and R_s than with MW, it is still an act of faith to assume that the gel filtration method is an accurate estimate of R_s for nonspherical species.

The use of SDS in PAGE is closely analogous to the use of guanidine hydro-chloride in gel filtration. The exact form of the relationship between R_f and MW or between K_R and MW for SDS denatured proteins has been studied (21). For instance, there is nearly a linear relationship between mobility and the logarithm of molecular weight, at least over a restricted range. This is analogous to the relationship between V_e/V_0 or K_{av} and log(MW) seen in gel filtration (23, 45). However, this linear relationship breaks down for both very large and very small proteins. If one assumes that the pore sizes obey a log normal distribution, then one can use a sigmoidal relationship between mobility and log(MW). Either a *logit* or *probit* type of analysis can be used: The logistic relations are given by:

$$R_f = a/(1 + (bMW)^c) \qquad\qquad (51)$$

or

$$K_{av} = 1/(1 + (bMW)^c) \qquad\qquad (52a)$$

or

$$V_e = V_0 + (V_t - V_0)/(1 + (bMW)^c)$$ (52b)

Alternatively, there is a nearly linear relationship between the logarithm of electrophoretic mobility and the arithmetic value of molecular weight (21). Indeed, a wide variety of linearizing transformations can be used from an empirical point of view. However, from a theoretical point of view, the relationship between logarithm of electrophoretic mobility and molecular weight seems most appropriate: a linear relationship is expected if all of the molecules have the same free mobility (e. g. after treatment with SDS), and if the surface area of the molecule is directly proportional to molecular weight (as would be expected for a random coil or for rod-shaped molecules (46)). Also, when using PAGE in SDS-buffers, one can use a linear relationship between K_R and MW (21). However, irrespective of the method of curve fitting employed, appropriate statistics should be used, such as the standard deviation of the estimate of the molecular weight estimates for unknowns, *including* the error in the estimation of the regression line (or curve). When this is done, then one finds that molecular weight determinations are generally far less precise than is usually imagined (e. g. (42)).

Even in the ultimate practical standard of reference, i. e. high speed sedimentation equilibrium analysis, it is necessary to use linear regression for estimation of the relationship between the logarithm of concentration and the square of the radius. One then finds the standard error and/or coefficient of variation of the slope. Thus, if there is a ± 7% error in the slope, there will be a ± 7% error in the estimate of molecular weight, if the partial specific volume of the protein is known perfectly. Since there may be a ± 10% error in the estimate of partial specific volume, there may be a ± 12% error in the estimate of molecular weight.

Accordingly, the results from PAGE must be compared with the competing methods (gel filtration, sedimentation, etc.) when the same type of statistical analysis and the same criteria are applied to each. The precision of the PAGE methods appears to be competitive with that of the other methods as usually practiced.

Electrophoresis of very highly charged molecules raises additional problems. Fisher and Dingman (figure 5 of (47)) have shown that rigid, double stranded nucleic acids have a nearly constant K_R value which is nearly independent of MW over a wide range, while a systematic, linear relationship between log K_R and log MW was obtained for flexible, single-stranded nucleic acids. This might indicate that these macromolecules break up the gel structure during their passage through the gel (a *ballistic effect*) (47, 52). The relationship between $\log(K_R)$ and $\log(MW)$ may be regarded as a maneuver to obtain linearization, and also to examine MW over a 1000-fold range. However, there is no theoretical basis for this relationship.

Free Mobility and Net Charge

Y_0 can be translated to values of molecular free electrophoretic mobility, M_0, and net charge, V (13). However, the conversion of R_f into an absolute mobility value entails considerable error. For example, in our standard procedure, it is necessary to know the relative mobility (RM) of the π-λ boundary (7), and the absolute mobility of the sodium ion under the same conditions of temperature and ionic strength:

$$
\begin{aligned}
M &= R_f\, \mu_f \\
&= R_f\, RM(1,9)\, \mu_{Na^+}
\end{aligned}
\tag{53}
$$

where RM(1,9) indicates the relative mobility of constituent 1 (the trailing ion) in phase 9 (the operative resolving phase, or π phase), and μ_{Na^+} represents the mobility of Na^+ at the same temperature and ionic strength. The mobility of sodium ion is well established. Of course, the conditions prevailing in the physical-chemistry laboratory for the measurement of μ_{Na^+} may differ from those in the typical laboratory performing gel electrophoresis. For example, the ionic strength may be seriously perturbed by the presence of ionic initiators used for the polymerization reaction. Also, the precision of the estimate of the displacement of the π-λ boundary is very poor (48). Even in the *Tris System* (based on that of Ornstein and Davis (11, 12)), there may be as much as a two-fold error in the (RM(1, 9)) and thus a two-fold error in the estimate of free mobility (and valence). It is possible to measure the boundary displacement (NU) directly (48), and compare this with the value predicted from the general theory of multiphasic buffer systems (7). However, the RM(1,9), the voltage gradient in the gel, and the mobility of the protein in the gel depend on the details of the preparation of the gel. The polymerization catalysts (persulfate, TEMED, etc.) can have marked effects on the ionic strength, pH and other properties of the gel (e. g. degree of polymerization (14)). To date, there is no satisfactory method for elimination of the effects of these catalysts (48). Gels can be purified from initiators, side-reaction products and reactants of the polymerization reaction in a manner similar to the method described by Morris (15). Such time-consuming pre-purification of gels by diffusion is not practical, at least for PAGE in MBS. This means that measurement of mobility (or free mobility) utilizing non-purified gels (or gels purified by pre-electrophoresis) are subject to potentially large systematic errors. Accordingly, when calculating either free mobility or net-charge, the results must be interpreted with extreme caution. Nevertheless,

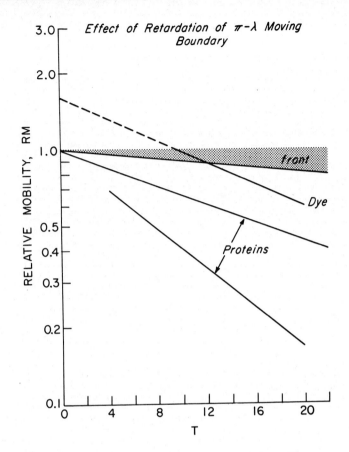

Fig. 6. Schematic representation of the effect of the gel on the apparent RM(1,9) and velocity of the front, and on the tracking dye. Here the tracking dye is unstacked when %T exceeds 12%. The relative mobility of the protein is first measured with respect to the dye, and then corrected by the R_f of the dye relative to the front by equ. 4, using the R_f for the dye on the basis of its Ferguson plot.

when one calculates the apparent free mobility of a protein at several pH values, using several different buffer systems, and then constructs a pH-mobility curve (figure 7 of (13)), the results are consistent with the pH-mobility curves obtained from free electrophoresis in the Tiselius apparatus, and/or from titration curves (e. g. (13)). Occasionally, one buffer system e. g. the one containing hydroxyethylmorpholine, (system C of (13)) will give anomalous results, with apparent mobilities differing by a factor of two compared to the other systems. This may be due to an error in the calculation of the RM(1,9), or to the possibility, that in this relatively hydrophobic solvent, many of the charged groups on the protein are folded inwards. Indeed, in virtually all buffer systems employed, the apparent charge on the molecule, calculated on the basis of the Debye-Hückel theory (49), is smaller, by about a factor or two, than the expected net-charge on the molecule if the molecule were completely unfolded, and if the pK's of the individual amino-acids were not altered by the protein structure.

Morris' method for measurement of M_0's (15) is independent of the problems of measurement of $(RM(1,9))$, μ_f, μ_{Na^+}, etc. However, it does not seem applicable to the use of multiphasic buffer systems, which are usually advantageous with respect to fractionation. His method is also free of the problems which arise due to the presence of the polymerization catalysts (KP, TEMED). However, both the methods of Morris and the present authors (13) are subject to the problem, that the estimated values of Y_0 or M_0 usually have at least a 10% error due to the extrapolation alone. This error can be reduced by increasing the number of points, by increasing the range of %T values, and by reducing the (weighted) average %T value. Also, both methods are subject to the problem, that the extrapolated Y_0 or M_0 appears to vary systematically with the crosslinking (%C) of the gel (26). Accordingly, it is impossible to know which one of the Y_0 values is correct. When one averages the results obtained at several gel concentrations, one obtains fairly good agreement with values obtained by free moving boundary electrophoresis (13, 15). However, results at different %C may vary as much as 50% (26).

Valence:

The calculation of net-charge (valence) involves so many assumptions, many of which can be shown to be erroneous before performing the studies, that it was rarely used even when data were available from free moving boundary electrophoresis studies. The assumptions and limitations of these calculations have been discussed in detail (13, 49). Calculations of net-charge must be interpreted in the light of these limitations: in absolute terms, the calculated net-charge may be erroneous by as much as a factor of two. Further, there is no practical way to validate or verify the results. Studies on a series of model compounds (such as the hemoglobin variants) will be necessary. Nevertheless, this type of calculation may be very useful, when studying a homologous series of proteins (e. g. the hemoglobins, the isohormones of human growth hormone (18), or an oligomeric series). Accordingly, the calculation of valence must be regarded as a research tool, and it is not ready for "routine", non-critical application.

Apparent Diffusion (Dispersion) Coefficient

In the ideal case, the spreading of the band in zone electrophoresis is governed only by diffusion. Thus, in principle, it should be possible to measure the diffusion coefficient on the basis of PAGE studies. However, to date this goal has not been realized, although the methodologies used (e. g. (2)) have been very crude. With the use of continuous scanning of the gel during electrophoresis (3, 53), it should be possible to obtain numerous measurements of σ^2 (square of the standard deviation of the band width) as a function of time and obtain much more reliable estimates (50, 3). One can calculate the apparent diffusion coefficient by

$$D = 1/2\ d(\sigma^2)/dt \qquad\qquad (54)$$

This diffusion coefficient must be extrapolated to zero gel concentration (%T), zero percent crosslinking (%C), zero sample concentration, and zero current (Joule heating). Corrections are then made for viscosity and temperature. In addition, corrections may be necessary for microheterogeneity of either the size or charge of the protein. To date, the observed band spreading would correspond in most cases to a diffusion coefficient between 10 and 100-fold larger than that predicted on the basis of the known Stokes' radius and molecular weight. This may be due to concentration dependence; or to microheterogeneity of the protein, heterogeneity of the gel, electrostatic interactions at low ionic-strength, etc.

One approach to the measurement of D by PAGE would be to use electrophoresis to introduce the sample into the gel in an ultra-thin starting zone (by stacking). Then, the current may be turned off, and the bandwidth measured as a function of time. This would be analogous to the use of the Tiselius apparatus for measuring diffusion coefficients, except that we are now dealing with zone electrophoresis rather than free moving boundary electrophoresis. Inherent in the use of this approach are the problems associated with the low ionic strength necessary for sufficiently rapid migration and minimal heating. The ionic strengths of electrophoresis are usually around 0.01 to 0.015. In contrast, in sedimentation and diffusion studies, the ionic strength is usually at least 10 times higher. Thus electrostatic effects become very severe. A similar problem arises, when one attempts to measure diffusion coefficients in isoelectric focusing, where the ionic strength is extremely low and also variable (3). In the approach just described, the ultra-thin starting zone is sandwiched between two different buffer phases, usually (though not necessarily) differing in pH and ionic strength, and necessarily differing in buffer ion composition.

Despite all of these problems, it is still useful to measure bandwith (width at one half maximum height, σ, etc.), to calculate the apparent diffusion coefficient (designated as dispersion coefficient, since true diffusion processes may be contributing only a small percentage of the total dispersion (2)), and to calculate the number of theoretical plates, N (equ. 47). These provide a measure of the resolving power and efficiency of the individual PAGE fractionation and, at least secondarily, provide some information concerning the nature of the molecular species undergoing fractionation, particularly with respect to homogeneity.

Optimization of Separation and Resolution

This is perhaps the most immediate reward of the use of quantitative PAGE, and justifies this approach even when fractionation, rather than characterization is the major purpose of the study. The rationales and procedures for optimization of pore size for both analytical scale and preparative *elution*-type PAGE are reported separately (30).

Conclusions

Quantitative PAGE allows one to utilize measurements of mobility of a molecule in gel electrophoresis at several gel concentrations to provide tests of homogeneity of a single preparation or identity of two preparations, estimates of molecular radius, weight, free mobility and net charge, apparent diffusion coefficient and number of theoretical plates.

References

(1)   Chrambach, A. and D. Rodbard, Science *172*, 440 (1971).
(2)   Lunney, J., A. Chrambach and D. Rodbard, Anal. Biochem. *40*, 158 (1971).
(3)   Catsimpoolas, N. and J. Wang, Anal. Biochem. *39*, 141 (1971); Catsimpoolas, N., Anuals N. Y. Acad. Sci. *209,* 65 (1973); Anal. Biochem. *54,* 66, 19, 18 (1973).
(4)   Ogston, A. G., Faraday Soc. Trans. *54*, 1754 (1958).
(5)   Rodbard, D. and A. Chrambach, Proc. Nat. Acad. Sci. *65*, 970 (1970).
(6)   Rodbard, D. and A. Chrambach, 3rd Interntl. Biophysics Congr., Boston, Mass., Aug. 1969, Abstr.
(7)   Jovin, T. M., Biochemistry *12*, 871, 879, 890 (1973).
(8)   Jovin, T. M., Annals N. Y. Acad. Sci., *209*, 477 (1973).
(9)   Jovin, T. M. and M. L. Dante, Multiphasic Buffer System Output, PB 196092, 1970. National Technical Information Service, Springield, Va.
(10) Jovin, T. M., M. L. Dante and A. Chrambach, Multiphasic Buffer System Output, PB No. 196085 to 196091, and PB No. 203016, 1970. National Technical Information Service, Springfield, Va.
(11) Ornstein, L., Ann. N. Y. Acad. Sci. *121*, 321 (1964).
(12) Davis, B. J., Ann. N. Y. Acad. Sci. *121*, 404 (1964).
(13) Rodbard, D. and A. Chrambach, Anal. Biochem. *40*, 95 (1971).

(14) Chrambach, A. and D. Rodbard, Sep. Sci. 7, 663 (1972).
(15) Morris, C. J. O. R. and P. Morris, Biochem. J. *124*, 517 (1971).
(16) Ferguson, K. A., Metabolism *13*, 985 (1964).
(17) Pastewka, J. V., R. A. Reed, A. T. Ness and A. C. Peacock, Anal. Biochem. *51*, 152 (1973).
(18) Chrambach, A., R. A. Yadley, M. Ben-David and D. Rodbard, Endocrinology, *93,* 848 (1973).
(19) McIlwaine, I., A. Chrambach and D. Rodbard, Anal. Biochem., *55,* 521 (1973).
(20) Corvol, P., A. Chrambach, D. Rodbard and W. Bardin, J. Biol. Chem. *246*, 3435 (1971).
(21) Ugel, A., A. Chrambach and D. Rodbard, Anal. Biochem. *43*, 410 (1971).
(22) Fawcett, J. S. and C. J. O. R. Morris, Sep. Sc i. *1*, 9 (1966).
(23) Rodbard, D. in preparation.
(24) White, M. L. and G. H. Dorion, J. Polymer Sci. *55*, 731 (1961).
(25) Neville, D. M., J. Biol. Chem. *246*, 6328 (1971).
(26) D. Rodbard, C. Levitov and A. Chrambach, Sep. Sci. 7, 705 (1972).
(27) Finney, D. J., Statistical Methods in Biological Assay, 2nd Ed., C. Griffin and Co., London 1964.
(28) Brownlee, K. A., Statistical Theory and Methodology in Science and Engineering, John Wiley, Inc., New York 1960, p. 314.
(29) Rodbard, D., G. Kapadia and A. Chrambach, Anal. Biochem. *40*, 135 (1971).
(30) Rodbard, D., A. Chrambach and G. H. Weiss, This volume.
(31) Yadley, R. A., D. Rodbard and A. Chrambach, Endocrinology, *93,* 866 (1973).
(32) Giddings, J. C., Sep. Sci. *3*, 181 (1969).
(33) Hedrick, J. L. and A. J. Smith, Arch. Biochem. Biophys. *126*, 154 (1968).
(34) Cantz, M., A. Chrambach and E. F. Neufeld, Biochem. Biophys. Res. Com. *39*, 936 (1970).
(35) Cantz, M., A. Chrambach, G. Bach and E. F. Neufeld, J. Biol. Chem. *247*, 5456 (1972).
(36) Myerowitz, R. L., A. Chrambach, D. Rodbard and J. B. Robbins, Anal. Biochem. *48*, 394 (1972).
(37) Ben-David, M., D. Rodbard, R. W. Bates, W. E. Bridson and A. Chrambach, J. Clin. Endocr., *36,* 135 (1973).
(38) Miyachi, Y., A. Chrambach, R. Mecklenburg and M. B. Lipsett, Endocrinology, *92*, 1725 (1973).
(39) Kaplan, G. N., R. D. Maffezzoli and A. Chrambach, J. Clin. Endocr. Met. *34*, 370 (1972).
(40) Maffezzoli, R. D., G. N. Kapalan und A. Chrambach, J. Clin. Endocr. Metab. *34*, 375 (1972).
(41) Sherins, R. J., J. L. Vaitukaitis and A. Chrambach, Endocrinology, *92*, 1135 (1973).
(42) Gospodorowicz, D., Endocrinology *90*, 1101 (1972); Banker, G. A. and C. Wilotman; J. Biol. Chem. *247*, 5856 (1973).
(43) Ryan, R. J., Biochemistry *8*, 495 (1969); Reichert, L. E. Jr., M. A. Rasco, D. N. Ward, G. D. Niswender and A. R. Midgley, Jr., J. Biol. Chem. *244*, 5110 (1969); Dufau, M. L., E. H. Charreau and K. J. Catt, J. Biol. Chem. *248*, 6973 (1973).
(44) Siegel, L. M. and K. Monty, Biochim Biophys. Acta *112*, 346 (1966).
(45) Andrews, P., Biochem. J. *96*, 595 (1965).
(46) Reynolds, J. A. and C. Tanford, J. Biol. Chem. *245*, 5161 (1970).
(47) Fisher, M. P. and C. W. Dingman, Biochemistry *10*, 1895 (1971).
(48) Chrambach, A., E. Hearing, J. Lunney and D. Rodbard, Sep. Sci. 7, 725 (1972).

(49)   Abramson, H. A., L. S. Moyer and M. H. Gorin, Electrophoresis of proteins and the Chemistry of Cell Surfaces, Hafner, New York 1942.

(50)   Ackers, G. K., J. Biol. Chem. *242*, 3237 (1967).

(51)   Pearson, E. S. and H. O. Hartley, Biometrika Tables for Statisticians, Vol. 1, 2nd Ed., Cambridge University Press, 1946, p. 57.

(52)   Dingman, C. W., T. Kakefuda, and M. P. Fisher, Anal. Biochem. *50*, 519 (1972).

(53)   Frederiksson, S., Anal. Biochem. *50*, 575 (1972).

Appendix I

Figure 6 shows the effect of the gel concentration on the velocity of the "front" and on the position of the dye with respect to the PI-LAMBDA boundary. For $\%T \leqslant 12$ (in this arbitrary example) the dye remains coincident with the front.

Acknowledgements

Dr. John Gart, Biometry Branch, National Cancer Institute, provided extremely useful advice and equations 42–44 for testing identity of slope and intercept simultaneously for two weighted regression lines.

2.3 Optimization of Resolution in Analytical and Preparative Polyacrylamide Gel Electrophoresis

D. Rodbard, A. Chrambach and G. H. Weiss

Abstract

Previous theoretical studies for optimization of analytical scale polyacrylamide gel electrophoresis (PAGE) are reviewed and extended. This expanded treatment permits computer simulation and optimization of the performance of both analytical (fixed duration) and preparative scale (fixed path-length elution) electrophoresis, at least under certain idealized conditions. Equations describing the concentration profile in the gel, the distribution of elution times or elution rate (concentration of protein in the eluate), and cumulative elution after any arbitrary time, have been developed. On the basis of the Ferguson plot, and the relation between the apparent diffusion coefficient and gel concentration, it is possible to optimize gel length and gel concentration to obtain maximal resolution per unit time. These computer simulation studies should make it possible to reduce the amount of experimentation needed before fractionation by preparative PAGE. The roles of several other factors are discussed, including pH, buffer constituents, stacking, ionic strength, voltage, surface area, load, viscosity, temperature, and elution buffer flow rate.

Introduction

Terminology: Analytical vs. Preparative PAGE.

For the sake of the present discussion, we shall use the term *analytical* poly-acrylamide gel electrophoresis (PAGE) to designate the situation where all components are subject to electrophoresis for the same length of time (i. e. fixed t). Then, the concentration profiles, c = c(x, t), or the centroids and moments around the centroid of each peak are analyzed (1).

The term *preparative scale PAGE,* refers to the case where all species must migrate a fixed length (L) and are then eluted (2). Thus, slower migrating species are subject to electrophoresis for a longer period of time (figure 1).

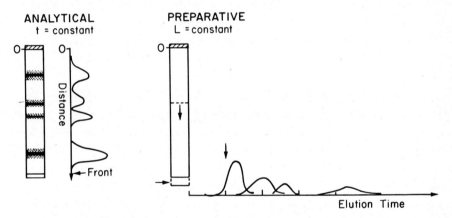

Fig. 1. Distinction between analytical (fixed duration, t) and preparative (fixed path length, L) electrophoresis.

These arbitrary definitions make no mention of surface area or load capacity. This is because, it is the choice between *fixed duration* or *fixed migration path length* which has the most important effect on procedures to optimize gel length and gel concentration. Perhaps these latter two terms are more appropriate, though less familiar.

Thus, a 6 mm diameter gel may be used for either *analytical* (3) or *preparative* (4) applications. Also, we can *prepare* protein preparations using the analytical constant-time gels, by slicing the gels and then removing the protein from the slices by diffusion, electrophoresis (5), or by dissolving the gel.

We shall give formal, mathematical definitions for separation and resolution below. However, we should comment that these are not synonymous: Separation refers to the distance between two bands, or the difference in elution times, without regard to bandwidth. Resolution refers to separation relative to the effective or average bandwidth, and thus provides (or can be converted to) a measure of the overlap between two peaks. Since the factors determining bandwidth are many and complex, expressions and results for resolution require numerous assumptions and approximations. By contrast, the results dealing with *separation* are quite exact, and only require validity of the Ferguson relationship (3, 6).

Choice of Buffer

Both the buffer system and the gel concentration must be optimized simultaneously to fully exploit the resolving power of PAGE. We shall first consider the role of the buffer separately. This is because, in practice, one must select the buffer system (pH, ionic strength, continuous or discontinuous) before examining the role of gel concentration. In subsequent sections, we shall examine the role of gel concentration both alone and together with choice of buffer.

The choice of a suitable buffer is particularly important for the fractionation of proteins, since it allows one to exploit even minor differences in molecular net charge for purposes of fractionation.

pH: In the optimization of pH, two conflicting considerations come into play. The closer the pH of the buffer to the isoelectric point, the greater is the likelihood of exploiting minor *charge* differences amoung proteins. In the extreme case, when the pH is between the pI's of the molecules being fractionated, the two species would migrate in opposite directions. However, the closer to the pI, the lower the net charge on the molecule, and accordingly, the lower the electrophoretic mobility. This, in turn, increases the duration of electrophoresis (for constant gel length, and voltage gradient), and thus the band-spreading due to diffusion. When multiphasic (18) (discontinuous (38)) buffer systems are used, one encounters an even more severe problem: as one approaches the pI, the protein of interest may fail to stack in the Upper (*Concentration* or *Stacking*) Gel. Usually, though not necessarily, the pH in the Stacking Gel is closer to neutrality than the Lower (*Resolving* or *Separation*) Gel. Another consideration in the choice of pH is that many proteins become relatively insoluble near their pI. Also, use of pH's which result in a mixture of polyanions and polycations (e. g. a mixture of protein[+] with acidmucopolysaccharide[−] or with nucleic acids[−]), may result in co-precipitation (coacervation)). This problem is aggravated at low ionic strength.

PAGE can be carried out at a pH anywhere between 2.5 and 11 in either a continuous buffer phase or in a multiphasic buffer system (MBS) (22). The pH of a continuous buffer phase is only limited by the high conductance and degree of Joule heating at the extremes of pH, and by the pH-stability properties of the protein. For example, deamidation and other hydrolytic reactions become severe at pH values below 3 or above 10.

In practice, it is desirable to start fractionation at an extreme of pH (e. g. 10), such that all — or at least most components migrate in the same direction. Then in subsequent experiments, the pH can be adjusted progressively closer to the pI. Also, in general, it is necessary to study a protein sample at both acid and alkaline pH, with cationic and anionic migrations, respectively.

Continuous versus Discontinuous (Multiphasic) Buffer System:

The use of a single, continuous buffer phase in PAGE has the advantage, that operative conditions are fully known and measureable. However, even here, the ionic polymerization catalysts KP and TEMED may seriously perturb the pH, ionic strength, conductivity, and voltage gradient. These effects cannot be removed, and may be aggravated by pre-electrophoresis. Although these effects can be eliminated, almost entirely, by prolonged washing of the gel in buffer, this introduces new problems in terms of swelling of the gel and alteration of the effective pore size (33). In continuous buffer systems, the velocity of a dye used as a reference for calculation of relative mobility values depends on the gel concentration (1). For Ferguson plot analysis (3) absolute mobilities have to be measured when a continuous buffer is used. But this is complicated by the fact that conductivity and voltage gradient are changed by the gel.

Continuous buffers also have the disadvantage that the thickness of the starting zone may become so large that it seriously impairs resolution. Therefore, relatively high concentrations of the protein sample are required.

Multiphasic (18) (or *discontinuous* (38)) buffer systems (MBS) have the major advantage that they permit fractionation of very dilute samples, since they can be used to form concentrated starting zones (*stacks*) prior to onset of fractionation in a Lower Gel. Also, the "front" serves as a convenient marker for measurement of relative mobility. These buffer systems have the disadvantage that the operative conditions (pH, ionic strength, buffer composition, etc.) differ from those under which the gel was polymerized. Although methods have been developed for measurement of the operative conditions in the gels with MBS, these are too complex and tedious for routine application (40). Fortunately, it is possible to predict the operative properties of MBS, on the basis of the comprehensive theory developed by T. M. Jovin (18 a, b). Several thousand buffer systems have been generated and can be retrieved by anyone from the magnetic tapes either by a listing of the output or by use of the available retrieval program (18 d). Further, the available programs (18 c) can be used for generation of additional systems, and for description of empirically derived systems.

From the viewpoint of fractionation, the most important property of MBS's is the generation of moving boundaries. These moving boundaries have the unique property of being able to concentrate electrolytes between the two buffer phases which constitute the leading and the trailing edges of the moving boundary, *when certain conditions are satisfied* (18). The concentrated electrolytes migrating between the leading and trailing constituent constitute a *stack* (38). All components of the stack migrate at the same velocity, which is regulated by the leading constituent (18). The ability to achieve deliberate concentration of either the protein of interest or of its contaminants within such a stack provides a tool for fractionation on the basis of mobility.

Before going into detail how one goes about using stacking for fractionation, it is necessary to define a number of terms and symbols of the theory of multiphasic buffer systems (see figure 1 of (18 a, Part II)). The homogeneous buffers on either side of a moving boundary are called *phases,* designated by greek letters (18 a, b) or by numbers (18 c, d):

Gel	(As formed, prior to electrophoresis)		Operative (After passage of moving boundary)	
	symbols	number	symbols	number
Upper Gel	BETA β	2	ZETA ζ	4
Lower Gel	GAMMA γ	3	PI π	9

After initiation of electrophoresis, and passage of the moving boundary, the BETA phase gel becomes a ZETA phase gel, and subsequently the GAMMA phase gel becomes the PI phase. The corresponding moving boundary is designated as the ZETA-BETA boundary or the PI-LAMBDA boundary. Each exhibits a characteristic *boundary displacement* and constituent mobilities of the buffer species on either side of the boundary. The numerical values for these mobilities are listed for each MBS in the computer output (18 d).

The leading and the trailing buffer constituents ahead and behind either the ZETA-BETA or the PI-LAMBDA boundaries are designated by numbers 2 and 1. The constituent mobilities of these two species are expressed relative to Na^+ at the same temperature and ionic strength, as relative mobilities or "RM". In the ZETA phase (designated as phase 4 in the computer program (18 d)), RM for buffer constituent 1 is termed "RM(1,4)". The value of RM(1,4) in each MBS defines the *lower stacking limit*. Similarly, RM(2,2)

defines the *upper stacking limit,* and RM(1,9) is the *unstacking limit* in the
Resolving Gel. In summary, the moving boundaries and their stacking limits
are:

	Stacking limits	
	Lower	Upper
BETA-ZETA	RM(1,4)	RM(2,2)
PI-LAMBDA	RM(1,9)	RM(2,8)

How are the values of RM(1,4), RM(2,2) and RM(1,9), listed in the output
(18 d) for each MBS, used for fractionation? How is stacking of the component of interest, or of its contaminants, in either the ZETA-BETA or the
PI-LAMBDA moving boundary of a MBS accomplished? How is stacking
made selective? This is done by the following type of experiment: A gel is
prepared in the buffer specified for either a BETA phase or a GAMMA phase.
The gel concentration is such that retardation of the protein is negligible
(e. g. a 3.125 %T, 20 %C gel). The protein mixture to be fractionated is
applied above the gel in the ZETA or BETA[1] phase buffer and subjected to
electrophoresis. The position of the ZETA-BETA, or the PI-LAMBDA
boundary, is marked either by a *tracking dye* concentrated in the boundary,
by measurement of the position of the inflection point in the sigmoidally
changing pH or conductance profile across the boundary, or by chemical
analysis for either one of the buffer constituents which disappears across the
boundary (3). The position of the stack in the gel may be recognizable as a
sharp refractive zone due to concentration of particular proteins or protein
fragments, even in the absence of a dye. The characteristic sharpness of a
stack is due to the fact that protein zones within a stack are not subject to
the ordinary degree of diffusion-spreading in proportion to time and migration distance (18).

Once the position of the moving boundary on the gel has been located by any
of these methods, the protein zones may be fixed and stained to reveal their
positions on the gel. Comparison of the protein positions with the position of
the moving boundary reveals whether their positions are coincident — the
"protein is stacked" — or whether the protein migrates either faster than or
more slowly than the stack. Since the values of RM(1,4) and RM(1,9) can be
varied systematically by simply using various buffers and buffer concentrations as shown in the MBS output, one can find buffers that will either stack

[1] When the leading ion (constituent 2) is a weak acid or base, the sample should be
applied in ZETA phase buffer; when it is a highly dissociating, strong acid or base,
the sample should be applied in the BETA phase (18). These general considerations
however may have to set aside in practice if the sample is insoluble or denatured at
the pH of the BETA phase. The Upper Buffer may be the ALPHA phase (any buffer
containing constituents 1 and 6 exclusively) or, arbitrarily, the ZETA phase (18 d).

or "*unstack*" the protein of interest, or any component of a mixture. This procedure is termed *selective stacking*.

Changing the constituents and composition of the buffer to vary the RM(1,4) and/or RM(1,9) is one way to achieve selective stacking. Another approach is to change the gel concentration while the buffer remains constant. In practice, one can alter the stacking limits, buffer system and the gel concentration simultaneously (see below).

Selective stacking can be used for the purpose of fractionation in one of two ways. Either the protein of interest is stacked and as many as possible of the contaminants are unstacked, or *vice versa*. Stacking of the component of interest has the advantages that the protein will be highly concentrated, that it will migrate without progressive band spreading due to diffusion and with high mobility (R_f = 1.0). These are obvious advantages in preparative PAGE. However, from the viewpoint of separation, confinement of the protein of interest to the stack has the disadvantage that the stack nearly always contains multiple zones of contaminants in addition to the protein of interest. These contaminants are migrating in separate but contiguous zones and cannot (except possibly under conditions of extreme overload) be physically separated from each other since they are only a few microns thick under the usual load conditions (18, 38). Thus, rather than selectively stacking the protein of interest, it may be advantageous to stack as many of the contaminants as possible, and to unstack the protein of interest by reduction of its constituent mobility below the lower stacking limit. This may be called *selective unstacking*. In preparative PAGE, this has the advantage that the highly concentrated stack sweeps the gel ahead of the protein of interest, and thereby saturates protein adsorbing sites in the gel which would otherwise lead to low recoveries (21).

Selective stacking or unstacking of the protein of interest by systematic variation of pH can be used for fractionation either in a single *Stacking Gel* (either phase ZETA or PI) or in a Stacking Gel (phase ZETA) superimposed on a Resolving Gel (phase PI).

The advantage of using separate, contiguous Stacking and Resolving Gels as in conventional *disc electrophoresis* is that the component of interest is concentrated in an ultra-thin zone in the Upper Gel before entering the Resolving Gel. Thus, almost unlimited volumes of very dilute samples may be applied to a given gel, without loss of resolution. The relative mobility of the trailing ion in the stacking phase, RM(1,4) of such a system can be increased or decreased systematically by change of pH to achieve maximum selectivity (although this has only rarely been used in practice). Alternatively, the gel concentration of the ZETA phase can be adjusted to provide selective stack-

ing. Similarly, the pH in the Resolving Gel (phase PI) can be altered systematically until the component of interest is selectively unstacked.

Selective unstacking can also be achieved while maintaining the operative pH
of the *stacking* phase constant throughout both an Upper and Lower Gel.
Gels are formed in either phase BETA or GAMMA; the sample is applied in
phase ZETA or BETA[1]. In this case, the protein of interest may be unstacked
in the *Resolving Gel* solely by virtue of an increase in gel concentration and
molecular sieving.

Summary: In order to use selective stacking (steady-state-stacking, isotachophoresis) for the protein of interest, a single gel of single buffer composition
(either the ZETA or the PI phase) and gel concentration suffices. If one
wishes to use *selective unstacking,* then it is necessary to use two gels in series.
Unstacking may be achieved by virtue of change in gel concentration, buffer
(pH, RM(1,4) or RM(1,9)), or both factors. By keeping the buffer constant,
one reduces the number of variables. However, in general, the buffer systems
have been optimized with respect to the "PI" phase; for example, the buffer-
ing capacity of the ZETA phase is often inferior in presently available systems
(18 d). Thus, when it is desired to use the same buffer in both Stacking and
Resolving gel, one should make both gels in the GAMMA phase, apply the
sample in BETA phase, and use the ZETA phase as the upper buffer.

Examples of the application of selective stacking to fractionations are the
purifications of murine and human α_1-antitrypsin (19) and of Hunter
Corrective Factor (20).

Ionic Strength $- I:$ The optimization of ionic strength is governed by the
same considerations as voltage gradient (see below), but several additional
factors come into play:

a) If I is too low, the buffer value of the buffer becomes inadequate.

b) The apparent diffusion coefficient, D, appears to be dependent on $I,$
   presumably due to suppression of electrostatic effects as I increases (23).
   (However, the data of (23) are not conclusive in this regard, since they
   involved a variation of time as well as of $I.$)

c) If ionic strength is increased, voltage gradient must be dropped precipitously to maintain constant current (or wattage); alternatively, the duration of electrophoresis must be increased or the length of the gel decreased.
   Conversely, lower ionic strength permits use of higher voltage gradients
   without an increase in Joule heating.

d) A reasonable compromise between the conflicting considerations of the
   effects of I on Joule heating and time of electrophoresis was adopted in
   the MBS output (18 d) which specifies $I = 0.015$ for the operative resolving

phase. Particular problems may require higher I; values up to 0.1 appear practical, although the prolonged time of electrophoresis at high I augments the risks of denaturation and the degree of band spreading.

I. A. Analytical PAGE: Maximum Separation

Equations for optimization of gel concentrations have been presented previously (6–8). We shall use the Ferguson relationship between mobility and gel concentration for the two species A and B in the form (3, 6):

$$\log_{10} R_f = \log_{10} Y_0 - K_R T$$

or,

$$M_A = A e^{-\alpha T} \qquad M_B = B e^{-\beta T} \tag{1}$$

$$\alpha = \ln(10)(K_R)_A \quad \beta = \ln(10)(K_R)_B$$

where

M_A, M_B = mobility of species A and B (Here, M is regarded as generic for any mobility measurement, either absolute or relative, e. g., R_f.)

K_R = retardation coefficient (3, 6)

= $- d(\log_{10} M)/dT$

T = gel concentration (corresponds to %T in previous papers; "%" omitted in equations for clarity).

We designate the Y_0's as A and B, and the retardation coefficients as α and β. The use of equ. 1 in this form avoids the need for frequent use of the term $\ln(10)$ to convert from natural to common logs.

Separation is simply defined as the difference in position after a fixed time (duration) of electrophoresis, i. e. the difference in mobilities (ΔM) multiplied by the electrical field strength and time:

$$\Delta x = \Delta M E t = (M_A - M_B)E t \tag{2}$$

$$= (A e^{-\alpha T} - B e^{-\beta T}) E t$$

where

x = position

E = electrical field strength

t = duration of electrophoresis

In some previous treatments (6–8) we have omitted the term E t, since this may be regarded as a constant for purposes of optimizing gel concentration, T.

Consider the relationship between mobility (M, or R_f) and gel concentration for two proteins (figure 2a). Let us assume that these two curves intersect when T = 0 (*size isomers* in the nomenclature of Hedrick (9)). Alternatively, assume that we have shifted or translated the horizontal scale so that $M_A = M_B = \mu$ when T = 0, and the mobility (or distance) is taken as unity at this point of intersection. At T = 0, there is no separation. As T increases, ΔM and Δx increase to a unique maximum, when $T = T_{max}$ (by definition). As T increases further, the mobilities of both species A and B asymptotically approach zero, so ΔM also approaches zero. Using only equation 2 (setting the first derivative with respect to T equal to zero), we find that maximal separation occurs when (6):

$$T_{max} = \frac{\ln(A\alpha)/(B\beta)}{\alpha - \beta} \qquad \alpha \neq \beta \qquad (3)$$
$$= 0 \qquad \alpha = \beta$$

When the two retardation coefficients are nearly (but not exactly) equal, so that $\alpha \cong \beta$, then we have several useful approximations (6–8):

$$\begin{aligned} T_{max} &\cong T_\mu + 1/\alpha \\ &\cong T_\mu + 1/\beta \\ &\cong T_\mu + 2/(\alpha + \beta) \end{aligned} \qquad (4)$$

where T_μ is the gel concentration at the point of intersection of the two Ferguson plots, i. e., when $M_A = M_B = \mu$, which is given by (6):

$$T_\mu = \frac{\ln A/B}{\alpha - \beta} \qquad \alpha \neq \beta \qquad (5)$$

At $T = T_{max}$, the average mobility (or distance) for species A and B will be reduced to approximately 0.368 = 1/e, (where e = 2.71828. . . = base of natural logarithms) multiplied by the mobility at the point of intersection (6), i. e.,

$$\bar{M} = (M_A M_B)^{1/2} \cong \mu/e = 0.368\mu \qquad (6)$$

Thus, the gel concentration giving maximal separation can be readily obtained by graphical means, once the two Ferguson plots have been obtained (figure 3).

Procedure:

1. Construct Ferguson plots (log R_f vs. T) for both species A and B.

2. Draw a line bisecting the angle between the Ferguson plots (dotted line). This represents the geometric or logarithmic mean of the mobilities of the two species, or $\bar{M}$.

3. Note the mobility at the point of intersection, μ.

4. Multiply μ by .368; draw a horizontal line at this level, and drop a perpendicular into the T-axis where this intersects the dotted line, $\bar{M}$.

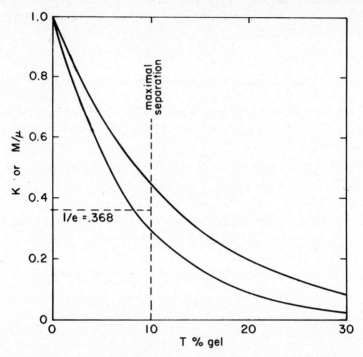

Fig. 2a. Mobility of two hypothetical proteins as a function of gel concentration. T_μ is arbitrarily taken as T = 0. Maximum separation occurs when the two mobilities have an average of μ/e. (Note: a similar relationship applies to K_{av} in gel filtration (6)).

5. Measure the difference, $\Delta M = M_A - M_B$ at this gel concentration.

6. Also measure the difference ΔM, when T = 0.

7. Compare the value of ΔM at T_{max} given in step 5 (from equ. 2 and 3) with the value of ΔM at T = 0 obtained in step 6. Select T_{max} or T = 0 depending on which gives the larger separation.

Comment: When the Ferguson plots are exactly parallel, so that $\alpha = \beta$, then equations 3, 4, and 5 do not apply. T_{max} occurs when T = 0 in this case.

Size Separation vs. Charge Separation:

One can set up an arbitrary classification scheme, to decide whether size and charge separation are antagonistic or synergistic (figure 4). This scheme is extremely simple, but very useful in practice.

Case A, with $T_\mu = 0$, corresponds to what Hedrick designated as "size isomerism" (9). This situation is frequently encountered with SDS denatured proteins (10, 11), and also with nucleic acids.

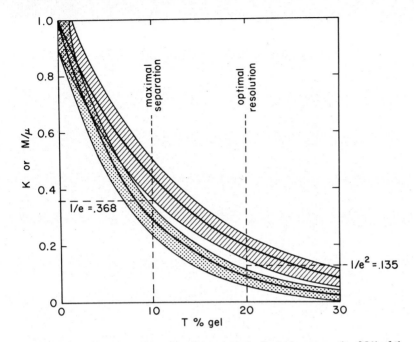

Fig. 2b. The centroid of each peak, together with ± 2σ limits (encompassing 95% of the protein) is shown as a function of gel concentration. As gel concentration increases beyond T_{max}, the apparent diffusion coefficient and the bandwidth decrease. Under idealized and simplified conditions, resolution is optimal when the mobility is reduced to μ/e^2, when T = T_{opt}.

Case B, $T_\mu < 0$, or $(\alpha - \beta)(A - B) < 0$: The molecule with the higher free mobility has the lower retardation coefficient (smaller size or radius). Thus, *size* and *charge* fractionation are *synergistic*.

Case C, $T_\mu > 0$, or $(\alpha - \beta)(A - B) > 0$: The larger molecule (steeper slope) has the higher free mobility. Here, size and charge separation are *antagonistic*. Unfortunately, this case occurs frequently in practice, since larger molecules usually have a higher net-charge and free mobility (if the surface charge density remains constant) (13). There is a positive correlation between Y_0 and K_R for a wide variety of proteins identified in sera (12). This case is particularly apt to occur when one is dealing with dimers, trimers, or higher oligomers of a common subunit (figure 4 of (14)).

When one is dealing with two distinct proteins, a change of pH may alleviate this size-charge antagonism. However, for an oligomeric series, with similar charge/mass rations, a change in pH is unlikely to improve the situation very much.

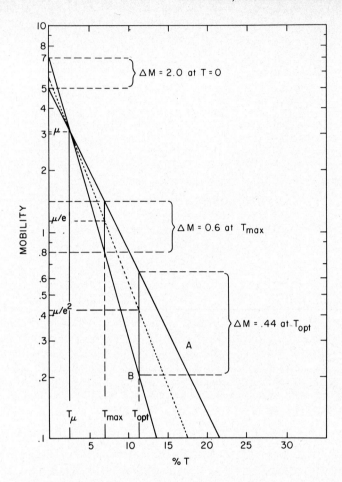

Fig. 3. Graphical method for locating T_μ, T_{max}, and T_{opt}. Ordinate: R_f or M on a logarithmic scale. Abscissa: Gel concentration. Procedure: 1. Draw both Ferguson plots. 2. Draw a line bisecting the two Ferguson plots. 3. Locate μ. Calculate 0.36μ and 0.135μ. 4. The corresponding gel concentrations are T_μ, T_{max}, and T_{opt}. 5. Compare separation and resolution at $T = 0$, T_{max}, T_{opt}.

Many times this antagonism of size and charge separation will occur over the entire available pH range. Since, in this case, ΔM and resolution are usually optimal at $T = 0$, "zero %T gels" are important, and this importance should stimulate the development of non-sieving but anti-convective media for gel electrophoresis (in addition to high %C gels, agarose gels, agarose-polyacryl-amide co-polymer, etc.), and development of other methods of *charge*

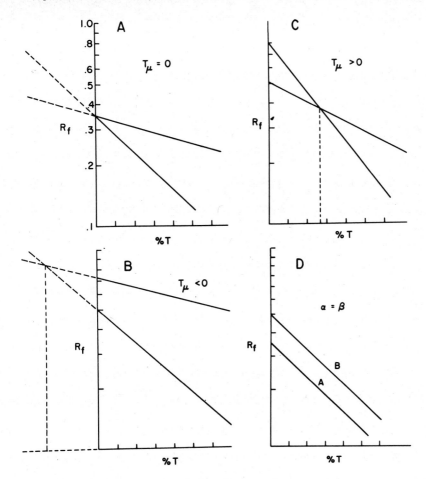

Fig. 4. A Classification of Separation Problems.

a) $T_\mu = 0$: "Size Isomerism" (9). This case frequently seen with SDS-PAGE.

b) $T_\mu < 0$: Size and charge fractionation are synergistic.

c) $T_\mu > 0$; Size and charge fractionation are antagonistic. This is the most frequent case. Seen with dimers, oligomers, etc. In general, larger proteins have higher free mobilities. One must consider $T = 0$ as well as T_{max} and T_{opt} in this case.

d) No T_μ; $\alpha = \beta$; "Charge Isomerism" (9). Here, $T_{max} = T_{opt} = 0$.

fractionation, e. g. isoelectric focusing, steady-state-stacking, and isotacho-phoresis (35).

Case D. T_μ nonexistent: $\alpha = \beta$. This is a case of *charge isomerism* (9) and is seen with molecules of the same size but different free mobility, e. g. hemoglobins (15) hGH isohormones (16), desamido species, etc. In this case, in

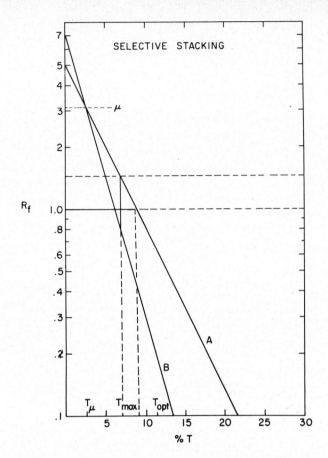

Fig. 5. Selective Stacking. The Ferguson plot may be used to optimize separation or resolution, subject to the constraint that the mobility of neither of the two species can exceed a given level ($R_f = 1$ or $M = RM(1,9)\ \mu_{Na^+}$). Also, the value of $RM(1,9)$ may be adjusted, to keep either the species of interest, or its contaminants, in the stack.

Fig. 5a. Here, $A = 5$, $B = 7$, $T_\mu \cong 3\%$, $\mu = 3.2$, $T_{max} \cong 7$, $T_{opt} \cong 11$. However, due to "stacking" when multiphasic buffer systems are used, the R_f of A is 1.0 at T_{max}; the R_f of B is ~ 0.80. By use of a 9% T gel, species A remains in the stack ($R_f = 1$), while the R_f of species B is reduced to approximately 0.42. Thus, a 9% T gel would provide better resolution and separation than $T = 0$, $T = T_{max}$, or $T = T_{opt}$. Use of a 13% T gel would *unstack* both species A and B. Both peaks would then be subject to diffusion spreading, and the resolution would not be expected to improve relative to that for a 9% T gel. If one could raise the stacking limit by a factor of 10, or if one were to use a continuous π-phase buffer, then a "0 % T" gel fractionation would provide greater separation ($\Delta R_f = 7 - 5 = 2$), but band thickness would be increased.

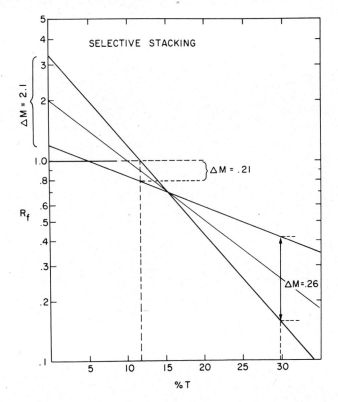

In the *Tris* systems (Buffer systems A or B of (3)), and in most other buffer systems
(18d), it is possible to increase the mobility of the trailing ion in the resolving phase −
the RM(1,9). This is shown schematically by the dashed horizontal line at R_f = 1.5.
In a "modified" system, this would become R_f = 1. This would make it possible to use
a 7%T gel, with species A remaining in the stack, while species B would have an R_f of
approximately $0.8/1.5 \cong 0.55$. This would permit use of T_{max}, and would be likely to
provide better resolution and higher maximal concentration of the peak, when compared
with a 9%T gel in the original buffer system.

Fig. 5b. Another example of selective stacking: A = 3.3, B = 1.2, T_μ = 15, $T_{max} \cong 30$.
Here the ΔM is largest at T = 0. At 12%T it is possible to "stack" A while B has an R_f
of 0.79. Although a larger ΔM could be obtained at T = 30, the stacking of A results in
a very narrow bandwidth, so that resolution at 12%T is superior. Note that gel concentration is a very critical parameter here. As %T changes from 12 to 15%, the resolution
decreases drastically. Likewise, at T ≤ 5%, there would be no resolution in the multiphasic buffer system shown here, since both components would be in the stack. Thus,
the moving boundaries of multiphasic zone electrophoresis introduce constraints which
may limit separation and resolution, but which also may be used to advantage to provide
improved fractionation. Indeed, use of the properties of the "discontinuous" buffer
systems may often override the calculation of T_{max} and T_{opt}. Inspection of the
Ferguson plots for the two species, makes it possible to evaluate those rather complex
considerations.

theory $T_{opt} = T_{max} = 0$. Accordingly, one should consider the use of isoelectric focusing in polyacrylamide gels (IFPA), isotachophoresis (17), cellulose-acetate electrophoresis, etc.

Selective Stacking: When using multiphasic buffer systems, it is *impossible* to have $R_f > 1$. Accordingly, if T_{max} occurs at a point such that the relative mobility, R_f, of either species A or B exceeds unity then we must modify our approach, using one of the following alternatives:

a) Draw a horizontal line at $R_f = 1$; then select the %T value that provides the greatest separation (or resolution, *v. i.*) subject to the constraint that both $(R_f)_A$ and $(R_f)_B$ must be less than or equal to 1.

b) Alter the buffer system to increase the value of RM(1,4) or RM(1,9) in the ZETA or PI phase respectively so that the species species of interest becomes unstacked (see above). This is described in detail in the Legend to figure 5.

c) Switch to a continuous buffer system identical with the ZETA, or PI phase respectively, of the multiphasic buffer system.

Ideally, one should work at a pH where size and charge separation are synergistic, i. e., the larger species has the lower free mobility. Otherwise, the effect of molecular sieving will reduce separation and resolution, so that free electrophoresis would be advantageous.

Preparative PAGE:

Although it may seem paradoxical, T_{max} is NOT necessarily the optimal gel concentration for analytical scale (constant duration) PAGE; this is due to the dependence of the apparent diffusion coefficient and bandwidth on gel concentration. However, as an approximation, T_{max} does provide the highest rate of resolution for preparative scale (elution) PAGE.

I. B. Optimal Resolution: Analytical PAGE

Figure 2b is identical with figure 2a, except that we have superimposed a shaded area corresponding to ± 2 standard deviations of the concentration vs. distance profile. Again, the ordinate may represent either M or R_f, and we have shifted the horizontal scale so that the two curves intersect when T = 0, i. e. we assume (momentarily) that $T_\mu = 0$.

Then, optimal resolution occurs when $T = T_{opt}$ which is distinctly greater than the T_{max} given in the previous section. By increasing the gel concentration above T_{max}, we reduce the separation between the centers of the two peaks, but we also reduce the band thickness, due to a decrease in the effective diffusion coefficient (perhaps better called the *dispersion* coefficient, since true diffusion is only a small component of the total band spreading) (23). This results in an improvement of resolution. However, if we increase the gel concentration too much, then the loss of separation will overcome the effect on band sharpness, and resolution is lost.

We wish to optimize Resolution, which we define as

$$Z = \frac{(M_A - M_B)\, E\, t}{(\sigma_A^2 + \sigma_B^2)^{1/2}} \tag{7}$$

where M_A, M_B are given by equ. 1, and σ_A^2 and σ_B^2 are given by

$$\sigma_A^2 \;=\; 2 D_A t \qquad\qquad \sigma_B^2 \;=\; .2 D_B t \tag{8}$$

$$D_A \;=\; D_A^0\, e^{-\alpha a T} \qquad\qquad D_B \;=\; D_B^0\, e^{-\beta a T} \tag{9}$$

$$D_A^0 \;=\; kT/(6\pi\eta R_A) \qquad D_B^0 \;=\; kT/(6\pi\eta R_B) \tag{10}$$

where

σ^2 = variance (2nd moment around the mean)
D = diffusion coefficient
D^0= "free" diffusion coefficient (in absence of the gel)
a = correction factor[2]
k = Boltzmann constant
T = absolute temperature
η = viscosity
R = molecular radius (Stokes radius)

These expressions for σ^2 and D assume: zero initial bandwidth, diffusion (or diffusion-like) bandspreading only, and that the correction factor, a, to relate diffusion coefficients to mobility as gel concentration changes, is the same for both protein species (1, 23, 24)[2]. Combining the above equations 1,7–10, we obtain:

$$Z = c_1 \frac{A\, e^{-\alpha T} - B\, e^{-\beta T}}{\left(\dfrac{e^{-\alpha a T}}{R_A} + \dfrac{e^{-\beta a T}}{R_B}\right)^{1/2}} \tag{11}$$

where

$$c_1 \;=\; E\, (t\, 6\pi\eta/(2kT))^{1/2}$$

Note 1 – This definition of resolution is independent of the total *mass* or concentrations of the two species. Other criteria for *resolution* may be necessary when there is a marked disparity in the total quantity (c_0) of the two species, depending on the objectives of the fractionation (25–30).

There is no universally accepted or adequate definition of resolution when both $\sigma_A \neq \sigma_B$ and the total mass of the two species is unequal, i. e. $(c_0)_A \neq (c_0)_B$. When $\sigma_A = \sigma_B$ and $(c_0)_A = (c_0)_B$, then almost all definitions of resolution lead to the same qualitative conclusions.

When $\sigma_A \neq \sigma_B$, most authors use the arithmetic mean σ:

$$\bar{\sigma} = \frac{\sigma_A + \sigma_B}{2}.$$

The denominator of equ. 7 is $\sqrt{2}$ multiplied by the quadratic mean σ. This approach has certain advantages:

a) When comparing two distributions, it is customary statistical practice to use the mean variance (σ^2), and not the arithmetic mean of the standard deviations.

b) The term $(\sigma_A^2 + \sigma_B^2)^{1/2}$ has another physical interpretation: this is the standard deviation of the distribution obtained as follows:

   1. Draw a molecule of A at random, using the concentration profile as a probability density function.

   2. Repeat for a molecule of B.

   3. Note the difference in position of A and B: $\Delta x = x_A - x_B$.

   4. Repeat steps 1–3 a *large* number of times, and construct the frequency distribution for Δx.

c) This method assigns a greater *weight* to the larger of the two σ's. This seems appropriate, since in general, the larger σ will be the limiting factor in obtaining resolution.

d) Finally, the method used here is analogous to the familiar Student's t test for difference between the means for two distributions.

Although the quadratic average increases the complexity of results slightly, this is no problem in practice, since computer or desk-top calculators are used for equ. 7. However, virtually the same results are obtained whether we use the arithmetic or the quadratic average.

Some authors multiply or divide the expressions for resolution by a constant factor (25–30). However, since resolution is on an arbitrary scale, this seems unnecessary.

When the mass of A is equal to the mass of B, and $\sigma_A = \sigma_B$, then the measure of resolution can be used to answer questions such as: "What percentage of A can be obtained with less than 1% contamination by species B?". Thus, the *purity*, percent contamination, or percent recovery can be calculated for any *cut point*. When $\sigma_A \neq \sigma_B$ and $(c_0)_A \neq (c_0)_B$, then all four of these parameters must be known (in addition to Δx) in order to make these calculations.

Note 2 — We discuss resolution in terms of two components. This is the simplest possible case, e. g., in the case of the protein of interest and its nearest contaminant, or when dealing with two species of interest. In actuality, there are often several *components* and several contaminants. In this case, we must apply the present calculations repeatedly, to find optimal separation between any two species, and it may be necessary to compromise conditions in order to obtain satisfactory separation between the species of interest and the two closest contaminants. Resolution of multicomponent systems is discussed in more detail elsewhere (12).

If we assume $\alpha \cong \beta$, $R_A \cong R_B$, $D_A \cong D_B$, and $a \cong 1$, then the expression for resolution is markedly simplified:

$$Z \cong c_2 \frac{Ae^{-\alpha T} - Be^{-\beta T}}{(2e^{-\alpha T})^{\frac{1}{2}}} \tag{12}$$

where

$$c_2 = c_1\sqrt{R}$$

By making a few simple, and usually plausible assumptions, we can greatly simplify the problem of finding T_{opt} in any particular case. We assume $\alpha \cong \beta$, i. e. that the retardation coefficients are similar, though not identical. After all, if the retardation coefficients were markedly different, then we have no problem whatsoever in obtaining adequate, if not optimum resolution. It is only when $\alpha \cong \beta$, as is often the case, that it is necessary to find T_{opt}. Further, we assume that $D_A^0 \cong D_B^0$, i. e., that the two species have similar diffusion coefficients in the absence of the gel. This follows, since if $\alpha \cong \beta$, the molecules are of nearly the same size, i. e., $R_A \cong R_B$.

Now, setting $dZ/dt = 0$, we find T_{opt} occurs (approximately) when (6):

$$\overline{M} = (M_A M_B)^{\frac{1}{2}} \cong \mu/e^2 = .135\mu \tag{13}$$

This will occur when (6):

$$T_{opt} \cong T_\mu + 2(T_{max} - T_\mu) \tag{14}$$
$$= 2T_{max} - T_\mu$$
$$\cong T_\mu + \frac{4}{\alpha + \beta}$$
$$\cong T_\mu + 2/\alpha$$
$$\cong T_\mu + 2/\beta$$

Thus, if we define T_μ as *zero*, (by translation of the horizontal axis), T_{opt} will occur when T is twice the value of T_{max}.

The estimation of an approximate T_{opt} proceeds along the same lines, and can be done on the same graph, as the estimation of T_{max}. Indeed, one need only

a) find T_μ
b) find T_{max}
c) find $T_{max} - T_\mu = \Delta T$
d) set $T_{opt} = T_\mu + 2\,\Delta T$

Alternatively, after steps 1–7 above (for finding T_{max}), continue with

Step 8) Multiply μ by $0.135 = 1/e^2$, draw a horizontal dashed line at this level, and find the gel concentration where this line intersects the line denoting $\overline{M}$.

Note: When $T_\mu \neq 0$, as is often the case in practice, then optimal resolution may also occur when $T = 0$. This is not given by the above equations, which fail to recognize a local maximum at $T = 0$, since they do not take into account that, in reality, T cannot assume negative values.

The resolution at zero gel concentration is given by

$$Z_{T=0} = \frac{(A - B)\,E\,\sqrt{t}}{(2D_A^0 + 2D_B^0)^{\frac{1}{2}}} \qquad (15)$$

where D_A^0 and D_B^0 represent apparent *free* diffusion coefficients. Unlike the comparison of separation at $T = 0$ and $T = T_{max}$, which is readily made by graphical methods, the comparison of resolution at $T = 0$ and $T = T_{opt}$ requires numerical evaluation. Fortunately, the choice between these two alternatives is readily made in practice by inspection of two band patterns, one obtained at T_{opt} and another using a gel with minimal molecular sieving (e. g., with $\geqslant 20\%$ crosslinking and low $\%T$). Of course, a true "0" $\%T$ gel cannot be made in practice; it is no longer a gel. However, sucrose gradient, paper, cellulose-acetate, and agarose zone electrophoresis come close to this ideal of "no" molecular sieving, although they introduce new problems (e. g. adsorption, endosmosis, etc.).

Computer Methods

The graphical techniques just described have proven to be very useful. However, we can obtain a more precise and accurate estimate of T_{max} and T_{opt} using computer methods. In particular, this enables us to consider cases where the assumptions that

$$\begin{aligned} A &\cong B \\ \alpha &\cong \beta \\ R_A &\cong R_B \end{aligned} \qquad (16)$$

$$D_A \cong D_B$$
$$D/D^0 \cong (M/M_0)^a, \text{ with } a \cong 1$$

do not apply.

The input parameters are:

1. Temperature (which influences the viscosity of the buffer, and thus the free diffusion coefficients and electrophoretic mobilities).

2. Duration of electrophoresis, t.

3. A, α, B, β (i. e. the Y_0's and K_R's) for the two species being separated.

4. R_A, R_B (which may be estimated from α, β).

5. The parameter "a" (6, 23, 24), if found to be different from unity using experimental methods as described by Lunney *et al* (23)[2].

Representative output of program "T-OPT" is shown in figure 6. Abscissa is gel concentration. Curve 1 shows the R_f (or M) for species A. Curves 2 and 3 show ± 2 standard deviations for the concentration-distance profile for this species, and the shaded area between these two lines shows the bandwidth containing 95% of that molecular species.

Curve 4 shows the position of the center of band B. Curves 5 and 6 show the ± 2 standard deviation region for this peak as a function of gel concentration.

Curve 7 shows the separation, ΔM, for these two species. Note a global maximum at T = 0, an abrupt fall to zero at the "μ point" ($T_\mu = 4$), and a rise to a maximum at $T_{max} = 13$, followed by a relatively slow decline (cf figure 4 of (31)).

Curve 8 shows the *average* band thickness. Actually, this is the standard deviation of the difference in position of a molecule of A and a molecule of B, when both molecules are drawn at random, as noted above.

$$\sigma_\Delta = (\sigma_A^2 + \sigma_B^2)^{1/2} = \sqrt{2}\, \bar{\sigma}. \tag{17}$$

This *average bandwidth* decreases with gel concentration, but at a rate about one-half that for the mobilities of the two species (when a = 1).

Finally, curve 9 shows the resolution. In this example, this also reaches its highest value at T = 0, falls to zero at T_μ, then rises to another maximum at T_{opt}, and finally declines again, but less rapidly than the fall in separation.

In general, T_μ is NOT zero, and T_{opt} is not always twice T_{max}. However, the relation given by equ. 14 still applies.

The maximal value for resolution calculated by this relatively exact method (without assumptions 16) may differ slightly from that given by the approxi-

---

[2]   The term a here corresponds to b in (6), a_3 of (23) or α of (24).

mations, equ. 13 and 14. The correct values for T_{opt} and for optimal Z could be obtained by numerical methods. The computer analysis automatically provides numerical comparison of the separation and resolution at T = 0, T_{max}, and T_{opt}.

Figure 6 displays R_f on a scale from zero to unity. However, the computer will re-plot the curves on any desired scale. This is useful, when dealing with continuous buffer systems, where mobility is not restricted to any arbitrary range.

We have discussed the problems and virtues introduced by the presence of a *stack* in the context of finding T_{max}. The same considerations apply, when

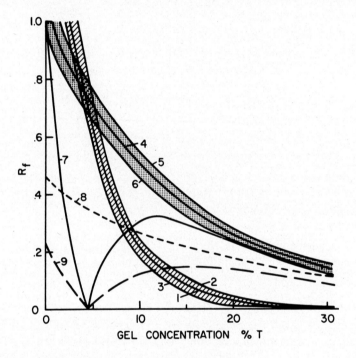

Fig. 6. Sample output from program T-OPT (re-drawn).
Ordinate:
1   centroid for peak A
2, 3   ± 2σ for peak A
4   centroid for peak B
5, 6   ± 2σ for peak B
7   separation
8   average band thickness
9   resolution
Note that the resolution is described by a very flat plateau region beginning near T_{max} and extending considerably beyond T_{opt}.

one seeks to optimize resolution. Only now, it is necessary to calculate the resolution vs. %T curves. For this purpose, the plot of the center of the bands, and the ± 2σ limits for each band is extremely useful (figure 6).

Discussion

From equ. 2, we see that separation increases linearly with time, although length of the gel must also increase proportionally. However, from equ. 11, resolution increases in proportion to $\sqrt{t}$. Thus, we would need a 100-fold increase in gel length to obtain a 10-fold increase in resolution. If there were microheterogeneity of the protein (42), then resolution incrases even less rapidly than $\sqrt{t}$. Thus, we run into a problem of diminishing returns. Also, the maximal protein concentration would be reduced by a factor of ten, thus straining the sensitivity of the method for detection of the protein. Note that both separation and resolution are proportional to field strength — until Joule heating becomes excessive.

Practical Constraints and Problems:

Sometimes the equations described here will give *impossible* values, e. g., negative values for T_{opt} or T_{max}. When this happens, then T = 0 (or the lowest possible practical gel concentration) should be used. Also, due to solubility properties of acrylamide and Bis, it is impractical to make gels with gel concentrations (%T) greater than 30–50%. If analysis indicates that the optimal *pore* size is smaller than that achievable by a 30–50% T gel, one can

a) adjust %C to 5%, giving the most restrictive gel (32–34).

b) use co-polymers, e. g. polyvinylpyrrolidone, to reduce the effective pore size for small molecular weight species (L. Ornstein, personal communication).

c) finally, one can increase the size of the molecules or reduce the effective pore size, e. g., by urea or possibly by use of denaturing detergents.

If T_{opt} or T_{max} is less than 3%, or if T = 0 is optimal, one can:

a) use polyacrylamide-agarose gels (35)

b) use 20–50% crosslinking (%C) gels (34, 41).

Use of any of these approaches requires construction of a new Ferguson plot.

If T_{opt} = 0, then one should again consider the use of isoelectric focusing, isotachophoresis in a non-restrictive medium, or electrophoresis on other media, such as cellulose acetate, agarose, agarose-acrylamide gels, or sucrose-density columns.

The preceding analysis is intended as a guide to the experimeter. However, the ultimate test is the experimental result. In a few instances, the above analyses indicated that T_{opt} and T_{max} = 0, (for parallel Ferguson plots for several species of rabbit homoglobins). However, in practice, the best resolution occurred when T = 8%; use of T = 4% gave serverely inferior resolution (15). This indicated that one of our basic assumptions was in error. In this case, it appears likely that this was due to a very strong dependence of D (and σ^2) on %T. Our methods for optimization would give the appropriate answer, if one were to use a = 2 in equ. 9 and 11. Also, we have ignored the effects of adsorption, molecular interactions, endosmosis, etc. on band dispersion (36). Finally, if there were severe heterogeneity of the protein preparation, e. g., a Gaussian distribution for A or α (Y_0 or K_R), or a Gaussian distribution of mobility (or R_f) for any specified gel concentration, then the apparent standard deviation of bandwidth may increase in proportion to time (42), i. e.,

$$\sigma \propto t$$

instead of

$$\sigma \propto \sqrt{t}$$

for the *diffusion only* or *diffusion-like* model used here. Accordingly, it is important to inspect the gel patterns. It is preferable to scan the gels by densitometer, and measure σ^2 as a function of %T for that particular protein. One can then use this empirical relationship as the basis for further optimization.

It is desirable to perform analytical PAGE under exactly the same conditions as to be used for preparative PAGE, as a preliminary to *scaling up*. Also, it is useful to analyze fractions obtained from preparative PAGE under the same conditions of mobility and gel concentration, to check for constancy of R_f and for unambiguous identification of components. This argues in favor of using the same gel concentration for both analytical and preparative PAGE. In practice, one can *compromise* and select a value of T intermediate between T_{opt} and T_{max}. As seen from the arbitrary example in figure 6, this intermediate value of T will result in only minor loss of *separation* or *resolution*, when compared to T_{max} and T_{opt}, respectively.

II. Preparative (Elution, Constant Path-Length) PAGE

The mathematical description of PAGE using *"prep"* elution columns is more involved than for the *analytical* case. Here we must optimize both gel concentration and gel length. Also, there are several additional problems and constraints. We have, until now, ignored problems of Joule heating and temperature gradients within the gel. This is not usually a major problem with small analytical scale gels. However, it is a major constraint in the design and optimization of apparatus and conditions (e. g. ionic strength, voltage gradient) for large preparative gels (2, 37). Also, we must choose between intermittent and continuous elution (21), and select either a constant or variable elution buffer flow rate. We also are dealing with a *time* constraint. In *fixed time* electrophoresis, we can readily detect a component with a mobility (R_f) only 0.05 relative to the front. However, this species will elute only when t = 20 (arbitrarily designating elution of the front as t = 1). Thus, we must wait a long time for elution of slow components. However, the likelihood of *accidents* (e. g., separation of the gel from the walls of the apparatus, jamming of fraction collector), and denaturation of the protein increase as time progresses. Also, the concentration of the protein decreases as time progresses. Accordingly, detection, reconcentration of fractions and/ or further processing of the fractions may become severe problems.

It is desirable to optimize resolution per unit of time, i. e., fractionation efficiency. Otherwise, we might be lead to a useless *optimum* requiring *infinite* time for elution.

Our *response variable* is no longer distance migrated; instead it is elution time, designated as t_e or t.

Finally, the problem of recovery or yield becomes of paramount importance, whereas usually this is of little (if any) concern for analytical studies (21).

We shall give, for didactic purposes, four progressively refined *approximations* to the solution of the problem of optimizing gel concentration and gel length in preparative PAGE. The first argues that the T_{max}, as defined above in the context of analytical gel electrophoresis, is a satisfactory approach to the problem of optimal resolution in preparative PAGE.

A second approach assumes that the distribution of elution times is approximately Gaussian, and governed only by diffusion or diffusion-like processes. The role of initial band thickness and of skewing of the peak is ignored.

A third approach assumes that the distribution of the protein concentration is Gaussian, but permits examination of the asymmetry of the distribution of elution times. A rough approximation is included to account for the effect of finite starting zone thickness.

Finally, the forced diffusion equation (Ficks Law) is used, to provide the exact concentration profile, whether starting from a finite or infinitesimally thick starting zone, and expressions are given for the distribution of elution time, the moments of this distribution, and the cumulative elution as a function of time, for the "zero" initial thickness case.

Also, the importance of considering resolution rate for preparative (elution) PAGE is discussed. These four approaches are used as the basis for computer simulation and optimization of the conditions of preparative-scale electrophoresis. The major limitation of all of these methods, is the difficulty of obtaining a satisfactory description of the relationship between band-width and gel concentration (23).

II. A. T_{max}

As an approximation, T_{max} as defined, calculated, and graphically estimated for the *analytical PAGE, fixed* t case above, provides the *best* gel concentration for preparative scale (constant path length or *fixed* L) PAGE.

The mean elution time (t, $\bar{t}$, or t_e) for the center of each peak is (approximately)

$$\bar{t}_A = L/v_A \qquad\qquad \bar{t}_B = L/v_B \qquad\qquad (18)$$

$$v_A = M_A E \qquad\qquad v_B = M_B E \qquad\qquad (19)$$

The standard deviation of bandwidth (in terms of position) is given by equ. 8. If we assume that A ≅ B (equal or nearly equal free mobility for both species), that a ≅ 1 in equ. 9, $D_A^0 \cong D_B^0$, and $\alpha \cong \beta$, then, as T varies

$$\bar{t} \propto 1/M \propto 1/D \qquad\qquad (20)$$

Thus, the effective band width at the time of elution is approximately constant for both proteins, and essentially independent of gel concentration.

$$\sigma_x^2 = 2\ D\ t \propto 2\ D(1/D) = constant \qquad\qquad (21)$$

The higher the gel concentration, the smaller the diffusion coefficient, but the longer the duration of electrophoresis prior to elution. These two effects will *cancel out if* a = 1 in equ. 9 and *if* bandspreading is due exclusively to these *diffusion-like* processes. Also, under the stated assumptions, and for fixed E and L, the bandwidth is nearly constant (either in terms of position or t_e) and insensitive to the retardation coefficients α and β. Accordingly, optimization of separation becomes (nearly) synonymous with optimization of resolution or resolving rate.

Alternatively, one could argue, that the maximal spatial resolution occurs when T = T_{opt}, but that the resolution is very nearly equal for all T values between T_{max} and T_{opt}. Then, use of T near T_{max} provides much more rapid elution, and thus better resolution per unit time. Reduction of T below T_{max}, towards T_μ results in a drastic reduction in resolution.

A more elaborate and exact analysis will be presented below. However, the use of T_{max} is a useful guide, will rarely be seriously in error, and permits the use of simple graphical analysis.

II. B. The Gaussian approximation: Zero Starting Zone Thickness

Ideally, the distribution of protein concentration and thus *apparent* velocity (i. e. position after finite t) is Gaussian (36, and below). However, the distribution of elution time is not perfectly Gaussian, since elution time is essentially the reciprocal of the *apparent* velocity. However, when the bands are narrow, and a variable x has a Gaussian distribution, then the variable 1/x is also nearly Gaussian with the same coefficient of variation, provided that the coefficient of variation of x or of 1/x is very small. This permits a description of the elution times in preparative PAGE which appears to be satisfactory for most purposes.

Assumptions:

1. Mobility and diffusion coefficients as per equ. 1, 8–10 (see above treatment of analytical gel case).

2. Initial starting zone thickness is infinitesimal compared with diffusion spreading. Actually, this is quite a reasonable assumption for PAGE with multiphasic buffer systems (MBS).

3. Distribution of elution times is Gaussian.

Since σ is a function of time, distribution of elution times does NOT follow the Gaussian distribution; instead it is (slightly) skewed. However, when the diffusion coefficient is small, and when the gel is *long*, this departure from *Normality* is not severe, and we may obtain several very useful results by means of a Gaussian approximation.

Let $\quad y \quad$ = distance migrated = vt $\hspace{5cm}$ (22)

$\quad \sigma_y$ = Standard deviation of position $y = \sqrt{2Dt}$

$\quad \overline{t}_e$ = Mean elution time = L/v

$\quad \sigma_{t_e}$ = Standard deviation of elution time $\cong (\sqrt{2D\overline{t}}/L)\,(L/v) = \sqrt{2DL/v^3}$

Note: When σ_y/y and σ_{t_e}/t_e are small (these terms having the same form of a coefficient of variation, CV), then

$$\sigma_y/y \cong \sigma_{t_e}/\bar{t}_e = \sqrt{2D\bar{t}}/(v\bar{t}) \tag{23}$$

$$= (\sqrt{2DL/v^3})/(L/v)$$

$$= \sqrt{2D/(Lv)}$$

By analogy with the Students t test, and with the definition of resolution used for the *fixed duration* electrophoresis, we may define resolution in terms of elution time as:

$$Z_t = \frac{(\bar{t}_A - \bar{t}_B)}{\sqrt{\sigma_{t_A}^2 + \sigma_{t_B}^2}} \tag{24}$$

where $\bar{t}_A$ and $\bar{t}_B$ represent the mean elution times for species A and B. (Again, we use the quadratic mean σ_t.)

If we combine equ. 1, 8–10, 18, 22, and 24, we obtain:

$$Z_t = \frac{\sqrt{LE}\left(\dfrac{e^{\alpha T}}{A} - \dfrac{e^{\beta T}}{B}\right)}{\sqrt{2D_A^0 e^{2\alpha aT} + 2D_B^0 e^{2\beta aT}}} \tag{25}$$

However, it may not immediately be clear from equ. 24 or 25 as to what is the optimal combination of T and L. However, numerical evaluation for a few combinations of L and T shows that the gel concentration for optimal resolution (as defined by equ. 25) is T = *infinity*. Thus, resolution as defined by equ. 25 is optimal when the RATIO rather than the difference, of mobilities is maximal. The Ferguson plots (log R_f vs. T) diverge indefinitely as %T increases, and the difference of the log R_f's is directly interconvertible with the ratio of mobilities. This ratio increases progressively as T increases:

$$M_A/M_B = A/B \; e^{(\beta - \alpha)T} = \bar{t}_B/\bar{t}_A. \tag{26}$$

Accordingly, despite their similarity to equ. 7, equ. 24 or equ. 25 are unsatisfactory as a measure of resolution, unless we optimize this function simultaneously with respect to T and L, subject to the constraint that the duration of the experiment is held constant at a finite value. For instance, we can keep $\bar{t}_A$ constant, by making the substitution that:

$$L = \bar{t}_A \, AE \, e^{-\alpha T}.$$

A similar equation is obtained if one wishes to maintain $\bar{t}_B$ constant. Depending on the purpose of the experiment, one might wish to keep $\bar{t}_A$, $\bar{t}_B$, or the average of these constant. However, in lieu of optimizing equ. 24 subject to this time constraint, one can optimize the resolution per unit time, given by:

$$RE = Z_t / t \tag{27}$$

The t in the denominator of equ. 27 may be $\bar{t}_A$, if one is primarily interested in species A; it may be $\bar{t}_B$ if one is primarily interested in species B; or it may be the average of these. Either the arithmetic, geometric, or harmonic mean of $\bar{t}_A$ and $\bar{t}_B$ may be used, with little or no effect on the optimal values for gel concentration and gel length. (One could also use the time corresponding to elution of any arbitrary percentage of A or B.)

Duration of electrophoresis: In analytical gel electrophoresis (non-elution), in the absence of protein microheterogeneity, resolution increases in direct proportion to the square root of elapsed time (all other conditions being held constant). Separation increases in direct proportion to time, while band width increases in proportion to the square root of time. Usually, the duration of electrophoresis is short enough, so that optimization of duration is no problem. Instead the choice of time depends on the length of the gel which can be conveniently handled (usually about 5 cm). *Time* is also determined by choice of voltage gradient which in turn depends on ionic strength and the maximal allowable wattage (heat dissipation) to maintain the temperature inside the gel within 1 to 2°C of the thermostat. However, for preparative scale electrophoresis, the choice of *time* is most intimately related to the selection of gel concentration and volume. In practice, elution times of approximately 5 hours have been effectively used in conjunction with elution buffer flowrates corresponding to 1 elution chamber volume per minute. Progressive deceleration of elution buffer flow (4a) may allow us to prolong the time of electrophoresis. The 5 hour period allows one to complete a preparative PAGE experiment on a single long workday, under continuous supervision and monitoring. Of course any change in temperature (0°C), I (0.01–0.015), current density (5–7 ma/cm^2), voltage (300–600 V), R_f-range (0.3–0.5) and elution flow rate will lead to different estimates of *optimal time*.

The above equations provide several insights:

1. Resolution increases in proportion to $\sqrt{L}$ or $\sqrt{t}$, as for the analytical scale problem (fixed time), whereas resolution per unit time is inversely related to $\sqrt{t}$ or $\sqrt{L}$. Thus, there are deminishing returns as we increase gel length.

2. Resolution increases in proportion to the square root of field strength, $\sqrt{E}$.

3. Using several representative numerical examples, the gel concentration which provides optimal resolution per unit time in equ. 26 was very nearly equal to T_{max} given by consideration of separation in analytical gels (equ. 3 or 4).

II. C. A modified Gaussian model

We now consider a slightly more refined approach to the use of the Gaussian approximation, to describe

1. the effect of finite starting zone thickness.
2. asymmetry of the distribution of elution time.

One simplistic treatment of starting zone thickness, is to assume a uniform distribution with a width of Δ at zero time. The variance of this distribution is

$$\sigma_0^2 = \Delta^2/12 \tag{28}$$

Since this initial variance is essentially independent of diffusion spreading, we can write

$$\sigma^2 = \sigma_0^2 + 2Dt \tag{29}$$

This amounts to introduction of a constant term in equ. 8. Further, we may assume that the overall distribution is very nearly Gaussian. Then, after a sufficient time, equ. 29 may be used for estimation of σ_A^2 and σ_B^2 in lieu of equ. 8 for numerical evaluation of resolution and resolution efficiency.

3. Asymmetry of c vs. t profiles can be investigated by caculating the time of emergence of the center of the peak and the time when any specified fraction of the total material has eluted. Let t_{A+}, t_A, t_{A-} denote the times of elution of 17%, 50% and 83% of species A (corresponding to + 1, 0, and - 1 σ of the concentration profile). Let t_{B+}, t_B and t_{B-} represent the equivalent elution times for species B. Thus, 66% of the protein will elute between t_{A+} and t_{A-} (figure 7).

Values for t_{A+}, t_{A-}, or t_{B+}, t_{B-} are obtained by solution of the quadratic equation:

$$L = vt \pm \sqrt{2Dt + \sigma_0^2} \tag{30}$$

or

$$at^2 + bt + c = 0$$

where

$$a = v^2$$
$$b = -(2D + 2Lv)$$
$$c = L^2 - \sigma_0^2$$

so

$$t = \frac{-b \mp \sqrt{b^2 - 4ac}}{2a}$$

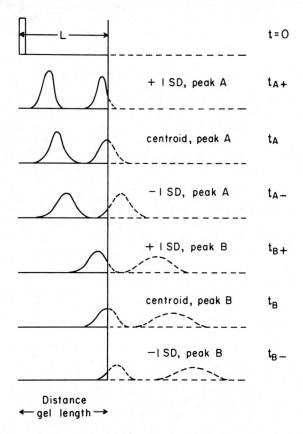

Fig. 7. Definition of various time points used to describe various stages in elution PAGE, using approach II. C.

t_{A+} and t_{A-} are symmetrical around

$$t = L/v \left(1 + \frac{D}{Lv}\right) \qquad (31)$$

Thus, the ratio of diffusion to the product of L and v is the key factor resulting in asymmetry of the distribution of elution time. Using equ. 30, we can calculate t_{A+}, t_{A-}, etc.

We may then re-define resolution as

$$Z = (t_A - t_B)/\sigma_{\Delta t} \qquad (32)$$

where various definitions for $\sigma_{\Delta t}$ may be used, e. g.:

$$\sigma_{\Delta t} = (t_{A-} - t_A) + (t_B - t_{B+}) \qquad (33)$$

or

$$\sigma_{\Delta t} = \sqrt{(t_{A-} - t_A)^2 + (t_B - t_{B+})^2} \tag{34}$$

The numerator is identical with that in our previous definition, equ. 24. The denominator is a measure of bandwidth, which includes the asymmetry of the leading and trailing edges, and the inequality of bandwidth for species A and B.

Alternatively, when there is serious overlap of the two bands, we may wish to collect only the leading edge of the first peak. Then our criterion for resolution might become

$$Z = t_{B+} - t_A. \tag{35}$$

In most cases, the conditions for optimal performance are little affected by this arbitrary choice of a definition or criterion for *resolution*, and are usually in excellent agreement with the *Gaussian* approximation, given above.

II. D. An exact expression for concentration profile, elution profile (flux, distribution of elution times), and cumulative elution

Expressions can be obtained for the concentration profile of a single non-interacting species in the column at any time, for the elution profile (equivalent to the distribution of elution times, or the concentration of protein in the elution buffer at any moment), and for the cumulative amount that has been eluted at any given time. These calculated profiles allow us to give a complete discussion of resolution, and enable one to calculate optimizing parameters.

We again assume that electrophoretic transport and longitudinal diffusion (or diffusion-like) bandspreading are the only operative physical factors. Then the development of concentration with time is describable by the Fick equation for concentration $c(x, t)$:

$$\frac{\partial c}{\partial t} = D \frac{\partial^2 c}{\partial x^2} - v \frac{\partial c}{\partial x} \tag{36}$$

where D is the diffusion constant and $v = ME$ is velocity. Both of these parameters are assumed constant, independent of position and concentration. If the sample is applied as a pulse of infinitesimal thickness (Dirac delta function) at $x = 0$, and the end of the gel is at $x = L$, the concentration pofile at time t is found to be (cf. figure 8 a):

$$\frac{c(x, t)}{c_0} = \frac{1}{\sqrt{4\pi Dt}} \left[\exp\left\{ -\frac{(x - vt)^2}{4Dt} \right\} - \exp\left\{ \frac{vL}{D} - \frac{(x - 2L - vt)^2}{4Dt} \right\} \right] \tag{37}$$

where c_0 represents the initial amount of the protein. Note: Due to the absence of molecular sieving in the elution chamber, and due to the continuous flow of elution buffer (in most apparatus), the protein concentration in the elution buffer (and at $x = L$) is effectively zero. Thus, entry of a molecule into the elution buffer is an irreversible step: it cannot diffuse backward into the gel. This boundary condition, viz. $c(L, t) = 0$, is responsible for the second negative exponential term on the right hand side of equ. 37. However, this has negligible effects in practice, as can be seen by evaluation of equ. 37 with representative parameters. If we remove this condition, we obtain a Gaussian distribution identical with that in section II. B.

The elution profile will be denoted by $J(t)$, so that $J(t)dt$ is the amount of protein to elute between t and $t + dt$. $J(t)$ is given by figure 8b:

$$J(t) = -D \frac{\partial c}{\partial x}\bigg|_{x=L} = \frac{c_0 L}{\sqrt{4\pi D t^3}} \exp\left\{-\frac{(L - vt)^2}{4Dt}\right\} \tag{38}$$

This function has a maximum at a time experimentally indistinguishable from $t = L/v$. Furthermore, the amount of protein eluted in time t can be written as (figure 8c):

$$P(t) = \int_0^t J(\tau)d\tau \cong c_0 \, \Psi \, (\sqrt{vL/(2D)} \, \ln(vt/L)), \tag{39}$$

where $\Psi(x)$ is the error function defined by:

$$\Psi(x) = \frac{1}{\sqrt{2\pi}} \int_{-\infty}^x \exp(-u^2/2)du. \tag{40}$$

It follows from equ. 39 that half of the protein is eluted at $t = L/v$. The parameters required for the calculation of resolution (equ. 25, 26) can be expressed in terms of the moments, μ_n, of the elution curve. These are defined to be:

$$\mu_n = \int_0^\infty t^n J(t)dt / \int_0^\infty J(t)dt. \tag{41}$$

The most important of these are:

$$\mu_1 = L/v \tag{42}$$

$$\sigma^2 = \mu_2 - \mu_1^2 = 2DL/v^3.$$

Thus, this (at least relatively) exact analysis confirms the important results obtained in the approximate treatment given in section II. B.

Finite Band Width: We shall now examine the effect of a non-zero initial band thickness, utilizing a more exact apprach than used in section II. C. For simplicity, we shall remove the boundary condition that $c(L, t) = 0$, since this was found to have negligible effect in the above case with infinitesimal initial band thickness.

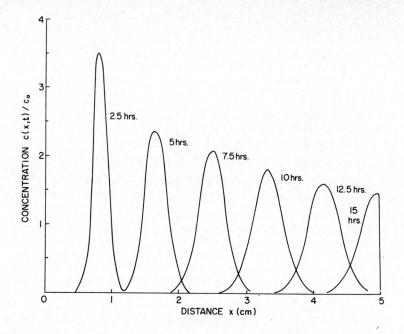

Fig. 8a

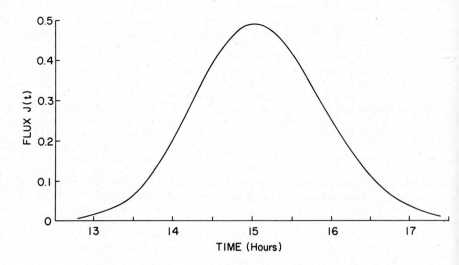

Fig. 8b

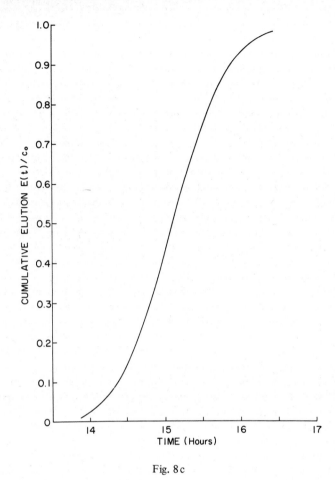

Fig. 8 c

Fig. 8. Computer simulation of the performance of a preparative PAGE fractionation.
Parameters used were:

A	=	2.5×10^{-5} cm^2 sec^{-1} volt^{-1}	D_A^0	=	3×10^{-6} cm^2 sec^{-1}
α	=	0.1	E	=	10 volts cm^{-1}
T	=	10	AE	=	2.5×10^{-4}
$e^{-\alpha T}$	=	e^{-1}	L	=	5 cm
a	=	1.5			

Note that D^0 is 10 times larger than the expected true diffusion coefficient.

a) Concentration profiles as a function of time. Initial thickness (equ. 37) is infinitesimal. Boundary effect at the elution chamber is exaggerated.

b) Flux of material into the elution chamber (equ. 38) (distribution of elution times).

c) Cumulative elution (equ. 39). This representation is useful for predicting cutpoints.

For finite zone width at $t = 0$:

$$c(x, 0)/c_0 = H(x) - H(x - \Delta) \tag{43}$$

(i. e. Δ = bandwidth at zero time; sample between $x = 0$ and $x = -\Delta$), one obtains

$$c(x, t)/c_0 = \Psi\left(\frac{\Delta - y}{\sqrt{2Dt}}\right) - \Psi\left(\frac{y}{\sqrt{2Dt}}\right) \tag{44}$$

where

$$y = x - vt.$$

Maximum peak height in the elution buffer:

The maximum peak height is of interest: if the protein becomes very dilute during the process of preparative PAGE, it will be necessary to employ an additional step to reconcentrate the sample; this in turn is likely to reduce recoveries. Equ. 38 permits us to calculate the maximum peak height.

For slower migrating proteins, the peak height is smaller than for the more rapidly moving proteins. Indeed, maximum peak height falls off precipitously as elution time increases: the band has been subjected to diffusion for a longer time, and it is moving more slowly (figure 9).

The maximal value of $c(L, t)$ for a peak is (from equ. 1, 9, 10, 37):

$$c(L, t_A) = c_0 \left(EAe^{aT}/(4\pi D_A^0 L)\right)^{\frac{1}{2}} \tag{45}$$

The maximal concentration in the buffer with fixed flow rate, q, is (from equ. 1, 9, 10, 38):

$$J(t_A) = c_0 \left(\frac{(EA)^3}{4\pi D_A^0 L} \exp\left\{-\alpha T(3 - a)\right\}\right)^{\frac{1}{2}} \tag{46}$$

At constant T, $J(t_A)$ is proportional to $1/\sqrt{L}$ and to $(EA)^{3/2}$. Thus a three-fold increase in either voltage gradient or free mobility results in a little more than a five-fold increase in peak concentration (it elutes sooner, the concentration at the end of the gel is higher, and it moves out faster). However, increasing gel concentration drastically reduces the maximum concentration obtained. Accordingly, it is desirable to use continuous deceleration of elution buffer flow rate (4) (figure 9). However, even with maximal deceleration (flow rate approximately inversely proportional to R_f or M), one still has progressive diminution of maximum peak height. Note: It is hazardous to reduce flow rate more rapidly than to have $q \propto 1/t$. When proteins reach the elution chamber, they are freed of the molecular sieving effect of the gel, and their velocity increases markedly, despite measures to increase the viscosity, alter the pH, or increase the ionic strength to reduce the voltage gradient in the elution chamber (2, Appendix II or III to 21). Thus, choice of buffer

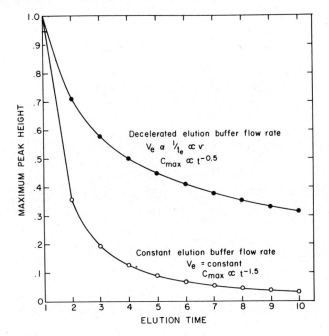

Fig. 9. Maximal concentration of protein in elution buffer, with
a) fixed elution buffer flow rate;
b) continuous deceleration of elution buffer flow rate, $q \propto 1/t$.

flow rate must take into consideration the free mobilities (A and B) of the
two molecular species, and flow rate must be rapid enough to prevent the
migration of the protein through the chamber and onto the limiting membrane or electrode.

III. Qualitative Discussion of other Factors

We have presented a series of models or approximations to permit optimization of concentration (T) and length (L) of the gel, to obtain the best
resolution for a given time period (or resolving efficiency per unit time) in
preparative (elution) PAGE. Simple graphical methods to find T_{max} will be
adequate in most cases. Computer simulation is necessary for the more exact
equations. Even the more complex models make a number of assumptions of
ideal behavior which are probably not realized in practice.

In addition to the optimization of gel concentration, the above considerations
govern the selection of the correlated parameters of time of electrophoresis

and gel length. We shall now briefly discuss several further parameters which affect performance of preparative PAGE:

1. Load/area
2. Cross-sectional area
3. Voltage gradient (current)
4. Temperature
5. Viscosity

Since these factors are highly interdependent, the optimization process is complex and results in more than one "optimum" set of conditions. There is no unique solution. Also the effects and interactions of these variables are more difficult to express quantitatively than the factors already considered.

1. Load/area: The load per surface area markedly affects resolution. The apparent diffusion coefficient increases with load (23); this effect "interacts" with ionic strength and/or duration of electrophoresis and possibly with gel concentration. Thus, the theoretical treatment given above, which assumes M and D independent of concentration, will need modification. Further, at extremely high concentration, the protein may perturb the pH, ionic strength, and voltage gradient within the zone (36, 38). Also, the load is one of the major factors affecting recovery: increasing load results in an increase in recovery, but a decrease in resolution (21). Empirically, it appears that 1–2 mg of the protein of interest can be applied per cm^2 of gel surface area, when the protein of interest is separated by one bandwidth from its nearest contaminant. This value was derived from separations of hemoglobins A and S, using 20–30 mg total hemoglobin/10 cm^2 of gel (2). When loads exceeded that range, separation between the band distributions of A and S remained incomplete under otherwise identical conditions. It should be emphasized that load is solely restricted to a low value for the proteins of interest and for components adjacent to that protein. The greater the separation between the contaminants and the species of interest, the greater the load capacity becomes. The total load of all components is nearly unlimited, except possibly by the amount of large molecular weight components that can be tolerated without clogging of the effective "pore" for the smaller molecular weight components. Thus, total load should not be used to characterize fractionation procedures and apparatus.

Recovery is directly related to load (21), apparently due to adsorption of proteins to polyacrylamide. Rather preliminary data on two proteins (21) indicate that, at a load of 0.1 mg protein/cm^2 of gel, recovery is 30%; at loads of 3.0 mg/cm^2 gel recovery increases to above 90%. This has been interpreted as an effect of adsorption of protein to polyacrylamide. Thus, it appears necessary for the purposes of both high resolution and high recovery in preparative PAGE, to limit the load of the protein of interest to 1 mg/cm^2, but to load contaminants at a level of 3 mg/cm^2 or more. Evidently, one will select "contaminants" with properties radically different from the protein of interest for this purpose, so that they can be removed easily subsequent to PAGE, e. g. by gel filtration. The "contaminants" should also have much higher mobilities than the protein of interest so that they can "sweep" the gel ahead of it by occupying adsorption sites (this will also have the beneficial effect of reducing the reactivity of stationary, uncharged free radical donors in the gel (39)). Choice of "contaminant" proteins is easy to make since it appears that all proteins, including those as small as bacitracin, are able to adsorb effectively to polyacrylamide.

2. Area: Presently available preparative PAGE apparatus provides too small a gel surface area to make it more than marginally milligram-preparative (2). Increasing the cross-sectional area of the gel results in a directly proportional increase in the applied load. However, this is accompanied by problems: if voltage gradient is to remain the same, there is also a directly proportional increase in current and Joule heating. This heat must be dissipated: however, the average path length to the surrounding heat-sinks is increased, and there is a rise in temperature within the gel. The annular geometry adopted by Jovin (2) for "milligram-preparative" size columns (up to 15 cm^2) circumvents this problem by providing a gel annulus thin enough (10–15 mm) to prevent a radial temperature gradient (21) when moderate voltages are used under "standard" conditions (Appendix III of (21)). Since at the voltages required for rapid fractionation the width of the gel ring cannot exceed 1.5 cm (due to the limited thermal conductivity of polyacrylamide), the volume of the elution chamber would triple if the surface area of the gel were doubled. As the diameter of the gel cylinder increases, the volume of the elution chamber becomes asymptotically proportional to the square of the gel cross-sectional area. Increase in the size of the elution chamber would cause more severe dilution of the eluted protein. Accordingly, the annular design of Jovin would need to be changed to increase the size beyond the present "maximum" of 15 cm^2. Larger gels also have relatively less wall adherence, due to increase in weight, thus necessitating better hydrostatic equilibration and mechanical stabilization.

3. Voltage: The higher the voltage gradient (or, for fixed ionic strength, current density in ma/cm^2 of gel) the shorter the duration of electrophoresis (for fixed gel length). This results in a decrease in diffusion spreading, σ^2 (equ. 8).

Elution time, $\bar{t}$, is inversely proportional to voltage, V. However, the higher the voltage, the higher the Joule heating:

$$W = Vi = i^2R = V^2/R$$
$$W = \text{watts}$$
$$i = \text{current}$$
$$R = \text{resistance}$$
$$V = E\,L$$

The greater the Joule heating, the smaller the size of the apparatus (and thus load) which can be run without marked deformation of the bands due to thermal gradients within the gel. It is apparent that optimization of voltage gradient requires simultaneous optimization of ionic strength, load capacity, temperature, duration of electrophoresis and gel concentration.

At $I = 0.015$, 0°C, a constant current can be maintained at 50–70 ma/ 15 cm^2 without elevating the temperature in the gel by more than a few degrees above the temperature of the thermostat in the course of a representative fractionation time of 5–10 hours (21). The degree of temperature sensitivity of the protein of interest as well as practical limitations on fractionation time and avoidance of band deformation have to be considered. Representative voltage gradients vary between 300 and 600 volts across all phases, and are estimated to be of the order of 10 volts/cm in the Resolving Gel. This usually corresponds to about 1 watt of power per vertical centimeter.

4. Temperature: The choice of temperature depends on the stability of the species under study. For enzymes, hormones, steroid-protein complexes and other active proteins, operation at 0°C is usually necessary. Otherwise, $25\,^\circ$C should be used for convenience, ease of polymerization, and generally higher electrophoretic mobilities by a factor of 2 (due to about a two-fold decrease in viscosity). The higher temperature is also accompanied by a slight increase in diffusion coefficient (see equ. 10). It appears that resolution generally improves as temperature increases. This effect is similar to that seen with increased voltage gradient i. e. it is due to shorter running time permitted by a decrease in viscosity.

5. Viscosity: Increasing viscosity *per se* results in a decrease in resolution. This may be of importance when sucrose is incorporated into the gel, or when performing electrophoresis in density-gradient stabilized columns. Also, certain buffers, notably Tris, when used in high concentration, can have a viscosity several times that of water or simple salt solutions (23).

Flowchart for preparative PAGE:

Despite the complexity and multiplicity of the considerations leading to optimization of the preparative-scale PAGE, we shall attempt to provide a pragmatic guide, as follows:

1. Select temperature, ionic strength, pH.

2. Select maximal duration of electrophoresis (e. g. 5 hours).

3. Obtain the ratio between the apparent and the true diffusion coefficient for your species of interest (23). In the absence of such data, assume that the apparent diffusion coefficient in the gel is 10 times larger than the true, free diffusion coefficient (23). Similarly, assume that the exponent a in equation 9 is 1.5 in the absence of empirical evidence to the contrary (this corresponds to some departure from the "ideal" case).

4. Measure K_R, Y_0, and $\overline{R}$ for both species A and B.

5. Select load, and thus, surface area, according to preparative requirements and availability of protein and apparatus. With knowledge of surface area, and the conductivity of the gel buffer(s), calculate the maximal allowable voltage gradient and current density.

6. Once the voltage gradient is specified, one can use computer programs based on the above equations to optimize %T and L, in order to provide the best resolution subject to the time constraint (item no. 2 above).

7. One can then, by use of another computer program based on the equations given in the preceding sections, calculate the predicted elution profile, and the resolution for any pair of species. These computer simulations are inspected, to determine whether performance can be expected to be satisfactory. These simulations are also helpful in determining *cut points,* at least approximately. These simulations also permit testing of a variety of conditions with little expenditure of time, effort, money or precious protein (figures 8 and 9).

Optimization of preparative PAGE conditions based on these considerations has been, and is being built into a series of computer programs. However, the existence of such programs does not diminish the importance to the experimenter of a thorough understanding of the principles and especially the limitations underlying the present approach.

References

(1)   Rodbard, D. and A. Chrambach, this volume, p. 28.
(2)   Jovin, T., A. Chrambach and M. A. Naughton, Anal. Biochem. *9*, 351 (1964).
(3)   Rodbard, D. and A. Chrambach, Anal. Biochem. *40*, 95 (1971).

(4a) Strauch, L. Protides Biol. Fluids Proc. Colloq. Bruges *15*, 535 (1967).

(4b) Ray, D. K., R. M. Troisi and H. P. Rapoport, Anal. Biochem. *32*, 322 (1969).

(4c) Vojtek, P., Anal. Biochem. *47*, 629 (1972).

(5) Ben-David, M. and A. Chrambach, Acta Endocrinologica, *72*, 654 (1973).

(6) Rodbard, D. and A. Chrambach, Proc. Nat. Acad. Sci. *65*, 970 (1970).

(7) Rodbard, D., G. Kapadia and A. Chrambach, Anal. Biochem. *40*, 135 (1971).

(8) Corvol, P., A. Chrambach, D. Rodbard and W. Bardin, J. Biol. Chem. *246*, 3435 (1971).

(9) Hedrick, J. L. and A. J. Smith, Arch. Biochem. Biophys. *126*, 154 (1968).

(10) Shapiro, A. L. and J. V. Maizel, Anal. Biochem. *20*, 505 (1969).

(11) Ugel, A. R., A. Chrambach and D. Rodbard, Anal. Biochem. *42*, 96 (1971).

(12) Kapadia, G., A. Chrambach and D. Rodbard, this volume, p. 115

(13) Abramson, H. A., L. S. Moyer and M. H. Gorin, Electrophoresis of proteins and the Chemistry of Cell Surfaces. Hafner, New York 1942.

(14) Thorun, W., Z. Klin. Chem. u. klin. Biochem. *9*, 3 (1971).

(15) McIlwaine, I., A. Chrambach and D. Rodbard, Anal. Biochem., *55*, 521 (1973).

(16a) Yadley, R. A., D. Rodbard and A. Chrambach, Endocrinology *93*, 866 (1973).

(16b) Chrambach, A., R. A. Yadley, M. Ben-David and D. Rodbard, Endocrinology *93*, 848 (1973).

(17) Chrambach, A., P. Doerr, G. R. Finlayson, L. E. M. Miles, R. Sherins and D. Rodbard, Ann. N. Y. Acad. Sci., *209*, 44 (1973.

(18a) Jovin, T. M., Biochemistry *12*, 871 (1973).

(18b) Jovin, T. M., Annals N. Y. Acad. Sci., *209*, 477 (1973).

(18c) Jovin, T. M. and M. L. Dante, Multiphasic Buffer System Output, PB 196092, National Technical Information Service, Springfield, Va.

(18d) Jovin, T. M., M. L. Dante and A. Chrambach, Multiphasic Buffer System Output, PB 196085 to 196091, 203016, National Technical Information Service, Springfield, Va., 1970.

(19) Myerowitz, R. L., A. Chrambach, D. Rodbard and J. B. Robbins, Anal. Biochem. *48*, 394 (1972).

(20) Cantz, M., A. Chrambach, G. Bach and E. Neufeld, J. Biol. Chem. *247*, 5456 (1972).

(21) Kapadia G. and A. Chrambach, Anal. Biochem. *48*, 90 (1972).

(22) Chrambach, A. and D. Rodbard, Science *172*, 440 (1971).

(23) Lunney, J., A. Chrambach and D. Rodbard, Anal. Biochem. *40*, 158 (1971).

(24) Weiss, G. H. and Rodbard, D., Sep. Sci. *7*, 217 (1972).

(25) Svensson, H., J. Chromatogr. *25*, 266 (1966).

(26) Vink, H., J. Chromatogr. *69*, 237 (1972).

(27) Rony, P. R., Sep. Sci. *3*, 357 (1968).

(27a) Rony, P. R., Sep. Sci. *5*, 121 (1970).

(28) Boyde, T. R. C., Sep. Sci. *6*, 771 (1971).

(29) de Clerk, K. and C. E. Cloete, Sep. Sci. *6*, 627 (1971).

(30) de Clerk, K. and T. S. Buys, Sep. Sci. *7*, 371 (1972).

(31) Richards, E. G., J. B. Coll and W. B. Gratzer, Anal. Biochem. *12*, 452 (1965).

(32) Fawcett, J. S. and C. J. O. R. Morris, Sep. Sci. *1*, 9 (1966).

(33) Morris, C. J. O. R. and P. Morris, Biochem. J. *124*, 517 (1971).

(34) Rodbard, D., C. Levitov and A. Chrambach, Sep. Sci. *7*, 705 (1972).

(35) Chrambach, A., G. Kapadia and M. Cantz, Sep. Sci. *7*, 785 (1972).

(36) Cann, J. R., Interacting Macromolecules, Acad. Press, New York 1970, pp. 14—19.

(37a) Hjerten, S., S. Jerstedt and A. Tiselius, Anal. Biochem. *27*, 108 (1969).

(37b) Brownstone, D., Anal. Biochem. *27*, 25 (1969).

(38)　Ornstein, L., Ann. N. Y. Acad. Sci. *121*, 321 (1964).
(39)　Dirksen, M. L. and A. Chrambach, Sep. Sci. *7*, 747 (1972).
(40)　Chrambach, A., E. Hearing, J. Lunney and D. Rodbard, Sep. Sci. *7*, 725 (1972).
(41)　Determann, H. and A. Walch, Z. Naturforsch. *27b*, 683 (1972).
(42)　Weiss, G. H. and D. Rodbard, Sep. Sci., in press.

2.4 Polyacrylamide Gel Electrophoresis with Discontinuous Buffers at a Constant pH

Robert C. Allen

The resolving power of PAGE systems is markedly improved when zone sharpening techniques are employed prior to or early in the separation phase of electrophoresis. Such sharpening may be achieved as described by Ornstein (1) and Davis (2) where sample components are sandwiched between a leading and trailing ion during a steady state stacking phase. While the basic electrochemical and operating conditions of this system were developed for the serum proteins, subsequent variations have been developed for other biological fluids based on the original theoretical considerations (3, 4). However, all of these systems require certain conditions to be met which may produce undesirable effects on the sample. If the sample is photopolymerized in the sample gel, the free radical formation resulting may reduce enzyme activity markedly (5). The discontinuities in pH may lead to undesirable effects on the sample also as discussed by Hjerten et. al. (6) or, on the other hand, the pH in the sample and stacking gel may be close to or approach the PI of one or more sample components leading to their not being included in the stack, or as in the case of LDH 5 in the original Ornstein system, they may not move out of the sample gel (7).

Similar resolving power may be obtained in discontinuous buffer systems at a continuous pH in conjunction with shifts in the conductance between the sample and separating gel as described by Hjerten et. al. (6), or by the passage of a moving boundary as described by Poulik (8). Allen et. al. (9), described a combination of these systems in which the sample placed in sucrose was first sharpened by a conductivity shift in a separating gel and then by a moving boundary which zone sharpened again the already separated components. However, the concentration potential of this system as routinely employed does not approach that of a steady state stacked system, thus, very dilute solutions may not be as readily resolved.

This report compares the above system with a modification which provides a more pronounced stacking effect on the already partially separated sample in a minimally sieving medium. In addition, the effects of various buffer systems, step gradient gel pore size, cross linking agent concentration (per cent C) and changing the location of stationary concentration boundaries and thus effecting stacking and unstacking are also compared.

Methods

A standard ORTEC 4200 electrophoresis system was used in this work modified only in that 1 mm thick gel slabs prepared in demountable cells were also employed[1]. A model 4100 pulsed power supply was utilized to provide either constant current or constant pulsed power. The flat slab system originally described by Allen, et. al. (9), designated system A was set up as shown in figure 1. The modified system designated system B was set up in the following manner.

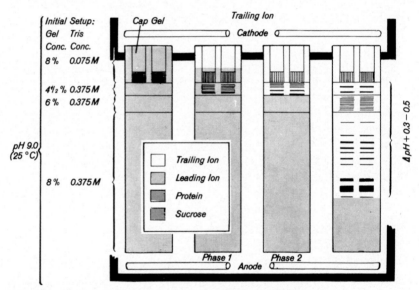

Fig. 1. Schematic presentation of polyacrylamide gel electrophoresis in step gradient gels with discontinuous voltage gradients.

A step gradient gel was prepared utilizing either by successive overlayering decreasing monomer concentrations with a syringe and canula from the top or by pumping in increasing monomer concentrations from the bottom utilizing a special base and system of syringes. A 48 mm column of 12% T gel overlayered by a 10 mm column of 8% T gel followed by a 5 mm column at

[1] Kindly supplied by OPTRO LTD., P.O. Box 2, Hornchurch, Essex, England.

6% T gel in 0.1875 molar tris buffer at pH 9.0 at 22°C (molarity in respect
to tris) was used. Next, a 10 mm column of 3% T gel in 0.0375 M tris buffer
pH 9.0 at 22°C was added and then water layered and allowed to polymerize
at 15–18°C. Gradients of 3%–8% for lipoproteins and 3%, 5%, 10%, 15%
and 20% for SDS systems were also prepared in a similar fashion. Leading
anions used in the tris buffer were chloride, sulfate, and citrate, while glycine
and borate were used as trailing ions. The cross-linking methylene bisacrylamide concentrations used for illustrative purposes were either 2.3% C or
3.5% C. The 3% T gel layer at the lower ionic strength was utilized to allow
the migration of larger macromolecules such as VLDL and LDL lipoproteins
and to allow more pronounced stacking to occur on the already separated
proteins following passage of the boundary, but prior to the entry of the
proteins into the 6% sieving medium. Thus, during stacking the slowest
sample component not sieved by the 3% T gel became the trailing ion. In
system B the lower ionic strength in the first gel layer and its larger pore size
allowed a longer migration prior to boundary sharpening and stacking as
compared to system A. In both systems A and B, well gel, cap gel, and sample
were at an ionic strength 1/5 that of the separating gel. Samples in buffered
sucrose or ethylene glycol were layered under the cap gel prior to cap gel
polymerization.

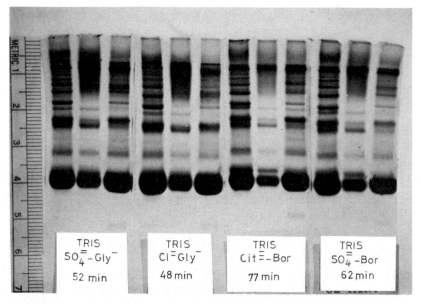

Fig. 2. Separation of normal and abnormal plasma in a 4.5–6–8% T 2.3% C step
gradient gels carried out at 90 ma constant current with 200 ugm of plasma protein
applied per sample. Gels are stained with Amido Black. System A.

Results

Replicate plasma protein pherograms utilizing 200 ugm plasma protein from
two patients with abnormal ceruloplasmin levels and a normal control were
separated at 75 ma constant current according to the method in system A.
The various leading and trailing ion combinations indicated marked differences in resolution particularly in the albumin and post-albumin region.
(Ceruloplasmin lies at 2.4 on the scale.) Comparing these results with the
migration ratios shown in table 1, indicated that in those systems where the

Table 1. Relation of boundary velocity to velocity of albumin front for various leading
trailing ion combinations at 75 ma constant current.

Trailing Ion	Leading Ion	Boundary Migration mm/min.	Albumin Front mm/min.	Ratio
Glycine	Chloride	1.91	0.79	2.17
Glycine	Sulfate	1.08	0.67	1.74
Glycine	Citrate	0.74	0.53	1.40
Borate	Chloride	1.07	0.71	1.51
Borate	Sulfate	0.95	0.62	1.53
Borate	Citrate	0.66	0.50	1.32

ratio of the boundary mobility to that of the albumin front is lowest, the
best resolution in the albumin and post-albumin region occurred. Where one
might expect diffusion to decrease resolution in the longer run times the
width of the albumin zone in the normal sample of each group is the same
or narrower in the citrate-borate and sulfate-borate systems, probably due to
the later unstacking in these systems.

A separation of replicate human plasma samples utilizing citrate as the leading
ion and borate as the trailing ion after method B is shown in figure 3. When
the same ionic strength buffer is utilized throughout the system, separations
A and C, a marked difference in resolution is apparent when compared to
separations B and D. This is due to the different stacking and unstacking
parameters caused by ionic strength differences when A and B and C and D
are compared, respectively. A marked change in the cathodal portion of the
gel occurs and the albumin zones in the single lower ionic strength gels are
much wider as might be expected. While the albumin, marked by the tracking
dye was allowed to migrate a fixed distance in each separation, the decreased
amount of cross-linking (%C) caused considerably more gel swelling during
staining and destaining, thus producing apparent longer migration paths
between the two systems.

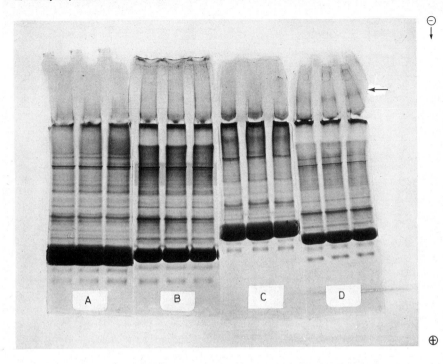

Fig. 3. Replicate samples of plasma separated with system B and modified B in a
3−6−8−12% T gel.

a) Buffer system tris-citrate-borate 0.0375 molar tris-citrate throughout the gel, 2.3% C.

b) Buffer system tris-citrate-borate with 0.0375 molar tris-citrate in sample well, cap
   and 3% T layer 0.1875 M tris-citrate in 6−8−12% T gel layers.

c) Similar gel buffer as in A but with 3.5% C.

d) Similar gel buffer as in B but with 3.5% C. Separation carried out with constant
   pulsed power and stained with coomassie brilliant blue R 250.

In figure 4, human plasma lipoproteins pre-stained with sudan IV in ethylene
glycol were separated in a citrate borate system after method B in a 3% T;
8% T step gradient gel. Here the VLDL (very low density lipoproteins)
chylomicron and pre-bata LDL (low density lipoproteins) Beta, and HDL
(high density lipoproteins) Alpha, are all well-separated with a least 4 distinct
alpha bands. When such separation is attempted in a single ionic strength gel
similar to A or C, figure 3, the pre-beta does not resolve and the beta appears
as a diffuse band only, again indicating the effect of stationary ionic strength
boundaries on stacking and unstacking and subsequent resolution.

A SDS system utilizing system B with sulfate as the leading ion and glycine
as the trailing ion in a 3%, 5%, 10%, 15%, 20% step gradient gel is shown in

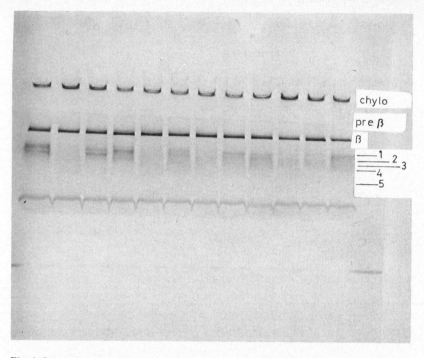

Fig. 4. Separation of plasma lipoproteins pre-stained with sudan IV in ethylene glycol. Carried out with system B citrate-borate in a 3%, 8% T gel 3.5% C. at constant pulsed power for 40 minutes. Albumin is stained with tracking dye.

figure 5, molecular weight species from 1450 (Bacitracin) to 67,000 (albumin) are readily unstacked in this system which employs a more rapidly moving boundary to prevent lower molecular weight species from moving with the boundary.

Discussion

The various parameters affecting the resolution of macromolecules in PAGE system utilizing discontinuous buffers have been discussed at length by Ornstein (1) and by Chrambach (10) and Rodbard (11) in this symposium. While this approach is predicated on stacking the sample components between a leading and trailing ion initially to zone sharpen, it can leed to certain problems in separation. First, it must be determined that all sample components will stack in the stacking gel under a given set of conditions. Thus, in a complex mixture of macromolecules, one or more of those present may fail to stack or migrate if the pH in the stacking phase is at or approaches

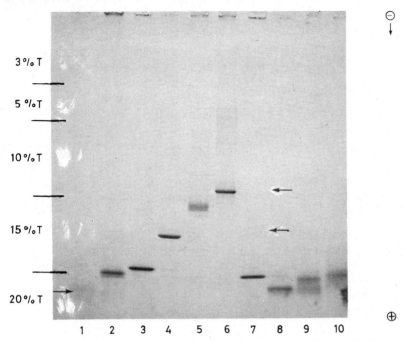

Fig. 5. SDS separation in 1 mm thick gel utilizing a 3−5−10−15−20% T 2.3% C. step gradient gel with method B in a Glycine-sulfate buffer system.
1 Bacitracin mw 1450, 2 cytochrome c   mw 12,400
3 Myoglobin mw 12,800, 4 chymotrypsin   mw 25,000
5 Ovalbumin mw 45,000, 6 albumin   mw 67,000
7 Hemoglobin mw polymer 68,000, 8 relaxin IV
9 Relaxin III, 10 Relaxin II

their PI. This is aptly illustrated in the original Ornstein system with LDH band 5 where the PI lies between pH 8.9 and 9.5 and thus, fails to stack or migrate out of the sample gel. A second consideration is the effect of poly-merizing samples in a gel where the resultant free radical formation may severely denature certain isoenzymes such as the esterases (5). Consideration must also be taken of the possible effect of pH shifts on the macromolecules present when stacking is carried out as has been discussed by Hjerten et. al. (6). Finally, one must also take cognizance of the possible protein-protein interactions as well as the possibility of ionic strength and pH causing co-precipitation under certain conditions during stacking.
The systems described here offer an alternative method to avoid the stacking step between the ion boundaries with its attendant pitfalls, yet provide equal resolution. This system has been utilized for both acidic and basic proteins and is amenable to a number of buffer systems at a variety of pH and ionic strengths (12). Since conventional stacking is avoided in this procedure,

electrophoresis is initiated at the running pH. For mammalian plasma and tissue extract systems a pH of 9.3 at 8°C has been chosen in an attempt to remain above the PI of the components present in the sample. However, in human plasma there is still a band, as yet unidentified, that moves in retrograde (see arrow figure 3d). However, since this system employs a cap gel of 6–8% T over the sample, this protein is readily visualized and may be quantified by microdensitometric methods.

The system described here is subject to the same unstacking effects as is the basic Ornstein system. Thus, more rapidly moving boundaries such as chloride-glycine and sulfate-glycine presumably unstack earlier thus leading to reduced resolution in the post-albumin region as compared to citrate-borate. This is in part overcome by utilizing step gradient gels where the post-albumin region, particularly the GC globulins, are resolved in 12% T gels rather than in 8% T gels (13).

On the other hand, rapidly moving boundaries and early unstacking appear to be desirable in the study of macromolecules denatured with SDS. In this instance, low molecular weight materials may not unstack and may travel with the boundary even in 20% T gels if the improper buffer system is chosen.

In all of these systems, it should be kept in mind that when gels or slabs are prepared by ammonium persulfate catalyzed polymerization that persulfate breakdown products, namely sulfate ions, are formed. This process will produce an additional boundary effect in systems utilizing ions other than sulfate as the leading ion. This may be readily demonstrated by reacting the gel with barium hydroxide or barium chloride. This produces and acid insoluble precipitate in the region of the sulfate ion and may be particularly misleading where boundary mobility or migration measurements are required in relation to a given protein.

The continuous pH systems at a discontinuous voltage gradient described in this report provide additional techniques for the separation of complex mixtures of macromolecules. The ability to produce single or multi-step gradients in slab gels also allows an increased flexibility for increasing resolution by both more selective unstacking and by step gradients to vary the sieving characteristics for maximal resolution of such complex mixtures. Gel thickness may be varied depending on the separation cell used from 1 mm to 1 cm depending on whether analytical or batch preparative techniques are desired.

Summary

A flat slab vertical PAGE system is described utilizing discontinuous buffers at a constant pH. The effects of various gel concentrations, buffer ions and ionic strength on resolution are compared.

Acknowledgements

The author wishes to express his deep appreciation to Dr. Bernhard Urbaschek, Institut für Hygiene and Medizinische Mikrobiologie der Universität Heidelberg, Klinikum Mannheim for the opportunity of working in this laboratory where system B was developed and to the Deutsche Kultusministerium for their generous support of the author during his tenure at Heidelberg. The author also wishes to acknowledge the expert technical assistance of Mrs. Rosemarie Hiemisch.

References

(1) Ornstein, L., Ann. N. Y. Acad. Sci. *121*, 321 (1964).

(2) Davis, B. J., Ann. N. Y. Acad. Sci. *121*, 404 (1964).

(3) Richards, E. G., J. A. Coll, and W. B. Gratzer, Anal. Biochem. *12*, 452 (1965).

(4) Jovin, T. M., M. L. Dante, and A. Chrambach, Multiphasic Buffer Systems Output. National Technical Information Service, Springfield, Va. PB 196092 (1970).

(5) Allen, R. C., R. A. Popp, and D. J. Moore, J. Histochem. Cytochem. *13*, 249 (1965).

(6) Hjerten, S., S. Jerstedt, and A. Tiselius, Anal. Biochem. *11*, 219 (1965).

(7) Disc electrophoresis newsletter *9*, Canalco Inc., Rockville, Md. (1967).

(8) Poulik, U., Nature *180*, 1477 (1957).

(9) Allen, R. C., D. J. Moore, and R. H. Dilworth, J. Histochem. Cytochem. *17*, 189 (1969).

(10) Kapadia, G., A. Chrambach, and D. Rodbard, this volume p. 115.

(11) Rodbard, D., A. Chrambach, and G. H. Weiss, this volume p. 62.

(12) ORTEC AN-32 A Techniques for High Resolution Electrophoresis. ORTEC Inc., 100 Midland Road, Oak Ridge, Tenn. 37830 (1973).

(13) Maurer, R., and R. C. Allen, Z. Klin. Chem. Klin. Biochem. *10*, 220 (1972).

Chapter 3. Evaluation of Polyacrylamide Gel Electrophoresis as Separation Method

3.1 Approaches of Macromolecular Mapping by Polyacrylamide Gel Electrophoresis

G. Kapadia, A. Chrambach and *D. Rodbard*

Introduction

The versatility of polyacrylamide gel electrophoresis (PAGE), obtained by use of variable pore sizes, pH and temperature, makes it ideally suited for the resolution of complex multicomponent systems (*mapping*). The concept of a *macromolecular map,* in analogy to the *fingerprints* of peptides, implies that it might be possible to detect minor pattern changes against the background of an otherwise complex pattern. This would make it possible to define characteristic changes in the band pattern of proteins, nucleic acids or polysaccharides for particular biological systems (tissues, organs, body fluids etc.) on their *macromolecular maps.* Such maps would be potentially useful as diagnostic tools for detection of stages of differentiation, disease states etc.

A great number of methodological approaches, using polyacrylamide gel electrophoresis (PAGE (1)) and isoelectric focusing on polyacrylamide (IFPA (2, 3)), have been previously applied to the resolution of multicomponent systems, particularly serum. However, the fractionation of multicomponent systems of proteins and other macromolecules, to display all components simultaneously, remains as an unsolved challenge, despite numerous recent advances in macromolecular fractionation methods. PAGE excels in resolving power compared to previous methods (1). Yet it has not been able to fully resolve a typical (and clinically most important) multicomponent system, human serum, by use of presently available techniques. The first serum patterns in PAGE (4, 5) had the immediate effect of "selling" PAGE as a *high resolution technique:* However, today it is necessary to also consider it as a *low resolution technique* in view of its failure to resolve serum into all or most of its components. Nonetheless, PAGE is still the most promising tool for macromolecular mapping (MM), and therefore the one we will consider exclusively in the present review.

Serum is known to contain 100–200 protein components (6): This poses problems of unambiguous identification of each species. Since components occur in serum in vastly differing concentrations. there are also serious problems of distribution overlap, with major components *overshadowing* minor components. What are the best ways to detect minor components in the presence of major ones? What are the best ways to spread components over the entire surface area of a two-dimensional *macromolecular map?* One should consider whether it is possible, desirable, or necessary to resolve serum by PAGE or other electrophoretic methods alone?

Aren't immunological tools, especially specific radioimmunoassays (RIA) far more sensitive tools for discrimination? Is it not better to have available a battery of specific tests for each of the serum components than a *macro-molecular map?* If the particular serum protein relevant to a physiological change is already identified, this will be undoubtedly the case. But MM can reveal the relative appearance or disappearance of one component in a complex system under varying physiological conditions even if we have no prior knowledge or hypothesis concerning the existence of the particular component.

Also, present techniques of clinical analysis, or other forms of analysis of multicomponent systems, focus on the analysis of the components that *can* be resolved rather than on those components that are necessarily medically or biologically important. A specific RIA will only detect changes with regard to one (or at most a few) molecular species. Similarly, from a functional view-point, the 30 stainable serum proteins discernible by PAGE are not likely to reveal the biological change of interest in any particular study. Yet, all present clinical applications of PAGE focus on these 30 proteins (or, just a few of these). In contrast a fully resolved macromolecular map should allow one to observe the *de novo* appearance or disappearance of a component on a map without any preconception concerning the nature of the changing component. Thus, MM should be capable of complementing the use of RIA and specific enzymatic assays in clinical practice.

An attempt was made to develop criteria and guidelines by which one may select the most efficient methodologies, among presently available techniques, and for purposes of evaluation and development of new techniques. This required, first, a survey of reported methods for MM, and the setting up of a classification scheme. Distinction was made between 1- and 2-dimensional (1-D and 2-D) or 2-stage methods, and those primarily responsive to mole-cular size or to charge.

Criteria for comparison of methods of MM:

1. The possibility to change the effective pore size in polyacrylamide gels continuously over a very wide range is a unique advantage for the resolu-tion of multicomponent systems involving a wide range of molecular

weights (1). This advantage is most efficiently exploited when electrophoretic fractionation is carried out in a pore-gradient (P-G-E) gel (7)
rather than a gel of medium size pore (4, 5, 8). Accordingly, the performance of various methods of P-G-E for the purpose of MM was compared by
the criterion of suitability for analysis of a wide molecular weight range.

2. Another criterion applied was that of fractionation efficiency, or resolving
   power, measured as number of theoretical plate equivalents, N (9)[1]. By this
   criterion 1-D methods appear severely limited, as do those 2-D methods
   that give rise to a clustering of components along the diagonal of the 2-D
   map. Maximal N, in contrast, can be obtained from 2-D (or 2-stage)
   fractionations which employ a different fractionation principle in the
   2 dimensions, e. g. utilizing mainly "charge fractionation" (10) in one and
   "size fractionation" (10) in the other dimension. However, 2-D maps do
   not readily permit one to define the characteristic positions of components
   quantitatively or with statistical confidence limits. Thus, they are subject
   to difficulties with regard to the interconvertibility of data between
   experiments and laboratories.

3. The mutual overlap of band distributions grossly differing in peak area is
   a general problem of MM. The greater the discrepancy in total protein
   mass, the more difficult it is to obtain satisfactory resolution (28). The
   degree to which various methods can overcome this problem, can be used
   as another criterion for the evaluation of MM methods. Overlap of
   component bands cannot be remedied in 1- or 2-D fractionation. Conceiv-
   ably, *2-stage* MM with analyses at the second stage at variable protein
   concentrations may be able to resolve minor components when they are
   adjacent to major ones.

4. The methods of MM can also be distinguished and evaluated by the
   frequency with which component pairs encounter the particular gel
   concentration (T_{opt}) required for their optimum resolution (7, 15, 31). By
   this criterion, the continuous pore gradient methods are favored over those
   using either invariant or discontinuously varying pore sizes. Of course, as
   the number of steps of the discontinuous pore gradient increases, its

[1] It has been suggested (45, 46) that the term and concept of "theoretical plates" be
abandoned. However, in its place, other measures of resolution or resolving capacity
are suggested which have a direct algebraic relationship (often a direct proportion-
ality) to the old concept of "theoretical plates". Thus, we may use the term
"theoretical plates" as generic for, and including, these closely related more refined
concepts of resolving capacity. The use of the classical concept of theoretical plates
has the advantage that it enables us to compare our results with those of other
fractionation methods (e. g. gel filtration, gas chromatography, liquid-liquid partition
chromatography, countercurrent distribution, sedimentation, etc.), albeit with some
reservations.

performance becomes progressively closer to that of the continuous pore gradient. For systems involving only a small number of components, the discontinuous pore gradient may be advantageous.

5. The degree to which various methods lend themselves to the numerical and statistical characterization of components on a *macromolecular map* represents another criterion for discrimination. Such characterization is presently feasible for 1-D fractionation and for 2-stage MM. Thus, 2-stage MM appeared promising and was evaluated, but results obtained indicate that between-experiment variability of R_f, K_R and Y_0 is too large to allow for the characterization of more than a small number of components.

6. A final criterion for discrimination between MM methods concerns the practicality of procedure and apparatus. Methods of MM differ widely in the availability, ease of assembly, ease of operation and disposability of apparatus. 1-D methods are superior with regard to these criteria. The available designs for 2-D gel slab apparatus are relatively difficult tu use since they generally require re-assembly and sealing of joints prior to each experiment. None of the existing slab apparatus have permitted the use of multiphasic buffer systems (MBS) (11) and sample concentration by "stacking" (4, 11) in the second dimension of fractionation. Most of the 2-D slab apparatus are made of plastic, which under many conditions provides inadequate adherence of the gel to the walls of the slab. In contrast, preparative apparatus which can be used for 2-stage MM is available and procedures are standardized (12), though cumbersome compared to procedures of analytical PAGE.

Macromolecular Mapping

Categories of MM:

Table 1 summarizes the various categories of MM, defines the abbreviations used for these categories and gives some examples from the literature for them. Distinction is made between one-dimensional (1-D) and two-dimensional (2-D) MM.

1-D Methods:

Among 1-D methods, the pore size can either be held constant, or it can be varied discontinuously (*step-function* gels) or continuously (pore size gradients). If a simultaneous analysis in both directions of polarity is desired, multiphasic buffer systems of both polarities can be devised that use 2 different separation gels of opposite polarity which are compatible with stacking gels of the identical buffer composition (11, 13). These are designated as *mirror systems;* potentially, these should be useful for the simultaneous mapping of polynucleotides and proteins. In addition to PAGE, 1-D fractionation may involve any one of the other *charge fractionation* (10) methods, e. g. isoelectric focusing (IFPA) (2, 3) or isotachophoresis (ITPPA) (14) on polyacrylamide gel, or electrophoresis on cellulose acetate or agarose gel.

2-D Methods:

There are 2 ways to conduct "2-D"-fractionations: either in 2 spatial dimensions on a gel slab, or in 2 stages in sequence, each involving a 1-D method. The latter approach is termed a "2-stage"-fractionation. Evidently the various 1-D methods can be applied in different combinations or sequences. Of course, one of the dimensions of fractionation need not be carried out on polyacrylamide; e. g. immunodiffusion or immunoelectrophoresis in the second dimension have been carried out on agarose gels (50). We have indicated the order of fractionation steps in 2-D MM by the order in which the abbreviations for the various 1-D fractionations are joined; e. g. ITPPA-PAGE stands for isotachophoresis on polyacrylamide in the first dimension of fractionation, followed by PAGE in the second. PAGE-ITPPA would indicate the reverse order of the two procedures.

The categories of 1-D and 2-D PAGE fractionations listed in Table 1 are not exhaustive, and are presented only to emphasize the vast array of available methodological combinations. We will now discuss the 6 criteria outlined above in greater detail and then apply them to the evaluation of the various 1-D and 2-D methods listed in Table 1, and will thereby attempt to indicate the methods of particular merit and promise.

Table 1. One-dimensional macromolecular mapping

Method	Abbreviation	Ref.[1]
Polyacrylamide gel electrophoresis	PAGE	
Extreme of pH, "medium" pore size		(4,5)
same, with SDS		(29, 30)
Simultaneously in cathodic and anodic direction		(11, 13)
Two or more pore sizes on superimposed ("step-function") gels		(25, 26)
Pore gradient electrophoresis	P-G-E	(7)
Electrophoresis at a right angle to the direction of gradient formation (transverse pore gradient)		(27, 44)
Isotachophoresis in polyacrylamide gel	ITPPA	(14)
Isoelectric focusing in polyacrylamide gel	IFPA	(2, 3)
Cellulose acetate electrophoresis		
Immunoelectrophoresis		

Table 1 (continued). Two-dimensional or two-stage macromolecular mapping methods

First dimension	Second dimension or stage	Reference[1]
PAGE	PAGE	
	(different pore size)	(32)
		see text
	(different pH and pore size)	(35)
	P-G-E	(36)
	transverse pore gradient	(Fig. 1)
	IFPA	
	immunoelectrophoresis	
	ITPPA	
P-G-E	PAGE	see text
	P-G-E	
	transverse pore gradient	
	IFPA	
	immunoelectrophoresis	
	ITPPA	
cellulose-acetate-electrophoresis	PAGE	(36)
	SDS-PAGE	
	P-G-E	
	immunoelectrophoresis	
IFPA	PAGE	(37, 39)
	SDS-PAGE	(40)
	P-G-E	(38)
	immunoelectrophoresis	
ITPPA	PAGE	see text
	SDS-PAGE	
	P-G-E	
	immunoelectrophoresis	

Note: Any of the above may be used either with or without multiphasic (discontinuous buffer systems (4, 11) in either the 1st or 2nd dimensions.

[1] Recent references are arbitrarily selected to serve as a guide to relevant literature.

Number of Theoretical Plates:

The major deficiency inherent to all of the methods proposed to date is the *finite* number of theoretical plates obtained. Giddings has shown (9) that the maximal number of detectable components or *peak capacity*, n (51), is directly proportional to the square root of the number of theoretical plates $(N)^1$.

$$n = 0.5 \sqrt{N}$$

Giddings has also provided formulas to calculate the number of theoretical plates for electrophoretic methods in the ideal case, when diffusion is the only source of band spreading. He calculates that an N of approximately 2000 can be achieved in the electrophoretic fractionation of most represent-ative proteins, using reasonable values for path length, voltage gradient and diffusion coefficient. The gel matrix and molecular sieving of the polyacryl-amide gel should result in a decrease in the diffusion coefficient (or apparent diffusion coefficient) compared with free solution. However, the electro-phoretic mobility is also reduced: in the *ideal* case, the mobility is reduced in direct proportion to the diffusion coefficient (17). This means that the use of a gel matrix can result in a loss of theoretical plates. In cases where the diffusion coefficient is reduced in proportion to the square of mobility (17), the number of theoretical plates is essentially independent of gel concentra-tion. In practice, the bandspreading of proteins in gel electrophoresis is one to two orders of magnitude greater than that predicted on the basis of diffusion alone. The numerous possible reasons for this have been discussed (17): heterogeneity of the gel matrix; micro-heterogeneity of the net charge or size of the protein; electrostatic effects, temperature gradients within the gel, concentration dependent effects, etc. Thus, in practice, it is possible to obtain between 200 and 2000 theoretical plates for an (analytical) polyacryl-amide gel 5 cm in length. Obviously, the number of theoretical plates can be increased by the simple expedient of increasing the length of the gel. How-ever, since resolution increases in proportion to the square root of N, one rapidly enters a region of diminishing returns. Also, as gel length increases, the duration of electrophoresis increases (proportionally), the problem of obtaining a uniform gel increases, and the problem of *detectability* of bands becomes accentuated, since maximum peak height is proportional to $1/\sqrt{t}$. However, gels of up to 50 cm have been used successfully (7).

Detection Sensitivity:

A criterion for evaluation for MM methods which is closely related to that of resolution is *detection sensitivity*. Even if a method could be found to distribute each of the protein components more or less uniformly along a line, or over a surface, each as a single sharp peak, non-overlapping with its neighbors, and with infinitesimal thickness (consider the output of a mass spectrometer), detection of all of the components would be difficult. Usually, staining methods (1) (followed by photography, densitometry, or just simple inspection) are used for analysis of results from electrophoretic fractionations. U. V. spectrophotometric scanning of proteins on gels during the course of electrophoresis has recently been introduced (18). These methods, however, have relatively low sensitivity. Use of both intrinsic and induced fluorescence detectors and fluorescent dyes (19) has been suggested and tried. However, to date these have not significantly reduced the detection limit for proteins in polyacrylamide gels. None of these methods provides a thousand- or million-fold range of sensitivity, which would be required to detect all of the proteins present in serum. Use of immunological methods (e. g. RIA) does provide this wide range of sensitivity (obtained by use of multiple dulutions). However, these kinds of assays are usually specific for one (or at most a few) components. Also, at the present time these methods cannot be applied to the entire, intact gel. Instead, RIA analyses require transverse slicing or progressive homogenization (1) of the gel: this results in serious loss of information. For example, if activity is found in two adjacent slices, it is difficult if not impossible to determine whether these represent one or two components; one is faced with the question: are the peaks and valleys real, or are they due to uneven slicing of the gel? Use of radioactive labels solves the problem of *detectability,* and also provides at least a thousand-fold range of detection (if one is willing to accept the radiation hazard or long counting times). However, usually, only one or two components are labeled. This can be used to advantage, when searching for a minor labeled component against a stained background. However, the use of radioactively labeled species usually involves the loss of information and resolution attendant to slicing of the gel. This may be avoided when using autoradiographs (1) of either longitudinally sliced gels or gel slabs. However, this reduces the range of detectability, since stray radiation from major components may overshadow adjacent minor components. The use of a labeled aminoacid (or nucleotide, etc.) to *globally* label all of the components in a mixture is one possible solution, but usually subject to a) non-uniformity of incorporation of label into different components, and b) inapplicability except to small *in vitro* and small *in vivo* systems. This is not a feasible approach for the mapping of components of human serum.

Thus, all of the methods for macromolecular mapping discussed so far can only be used for the detection of *major* components, which probably number fewer than 50 in human serum. In the ideal case, one could increase the load indefinitely, to bring the minor components to the level of detectability. In practice, this is impossible. With the exception of steady-state stacking (4, 11) (SSS) and isotachophoresis (ITP) (14), the resolution of all fractionation methods decreases with increasing load. Also, except for the ideal *mass-spec* situation, where diffusion (and other sources of peak broadening) are neglected, increasing the load of sample applied does not increase the ability to detect minor components, which remain as barely detectable *shoulders* or irregularities on the tails of the Gaussian distributions of major components. As noted above, SSS and ITP do permit (at least in theory) unlimited increases in load. This raises the intriguing possibility that these methods could be combined with others (in a two-stage procedure), so that the concentrations of major components could be reduced (by dilution in subsequent analyses), but the concentrations of minor components could be brought to levels of detectability. This approach assumes that the starting material is available in abundance; this is true for serum, but not true for many other cases (e. g. tissue-culture extracts). Another approach to solve this problem imposed by the limited detection limits of staining and related methods, is to combine a preparative-scale fractionation with analysis of the fractions by a relatively high resolution method. This has been tested, as discussed below.

Inherent in the detection problem, and related to the capacity of methods to spread components evenly over the surface of a map, is the mutual overlap of band distributions, particularly since components occur in vastly different concentrations in multicomponent systems. Thus, the detection of minor components is suppressed by the presence of major ones, since their band distributions are superimposed. Thus, barring the use of specific detection agents (e. g. RIA) all 1-D and 2-D procedures reveal only the most concentrated components, while 2-stage procedures may potentially be able to solve this problem.

Component identification:

For the purposes of recognizing a particular component within a complex pattern it is desirable not only to recognize the component (as on a peptide map) but also to identify it in some quantitative, numerical fashion within confidence limits. At any one gel concentration, such identification on a 1-D gel is readily carried out in terms of R_f and σ_{R_f} (16, 20). When studying a multicomponent system at several discrete gel concentrations (using different gels), it is very difficult, and often impossible, to know which bands correspond to the same component. Therefore, it is usually not possible to characterize such a component by the physically meaningful parameters K_R and Y_0 in PAGE (15). For example, of the 22 components of serum recognizable at pH 10.2 by R_f (Table 2), Ferguson plots can be constructed for only 11 of these components with certainty, using the methods previously discussed (20).

Optimal gel concentration:

Each pair of components is optimally separated at a mathematically defined gel concentration, the T_{opt} (15, 31). Methods can therefore be evaluated according to the fraction of component pairs being fractionated at, or close to, their T_{opt}, and/or according to the percentage of the fractionation time during which component pairs are exposed to their T_{opt} or a value of $\%T$ close to it.

Choice of methods responsive to molecular charge and size:

Size and charge fractionation (10) usually do not provide the same number
and sequence of components, and their successive application to 2-D MM
should therefore spread components away from the diagonal of a map.
Therefore, the fractionation method utilized for the second dimension of
2-D mapping should operate on a principle independent of that used for
separation in the first dimension. Ideally, the various species of a multicomponent mixture would be spread randomly over the two-dimensional
plane. For example, if the first dimension fractionated molecules on the basis
of net-charge (or free mobility), while the second dimension separated
molecules on the basis of size, one might expect that every region of the 2-D
gel slab would be covered with *spots*. This goal is very nearly achieved in
paper-electrophoresis/chromatography (*fingerprinting*) for peptides, where
molecules are fractionated on the basis of *charge* in one dimension and on the
basis of their partition coefficients in hydrophobic solvents in the second
dimension. Of course, this makes the implicit assumption that the distribution
of *net charge* or free mobility and the distribution of molecular size are
independent when considering the population of proteins in serum (or other
multicomponent systems). Actually, charge density and molecular radius are
not *independent* parameters. There is a correlation between the results of
pure charge fractionation and *pure size fractionation.* Larger molecules tend
to have a higher free mobility, since mobility is directly proportional to net
charge (and thus surface area, R^2), while inversely proportional to Stokes
radius (R_s), as a rough approximation (21). The methods which might be
expected to provide more-or-less pure *charge fractionation* would be zone
electrophoresis on paper, cellulose acetate, agarose gels, polyacrylamide gels
with $> 15\%$ crosslinking (22, 52) or with very low total gel concentration
(23), or the use of gels composed of both polyacrylamide and agarose (14).
In addition, Steady-State-Stacking (SSS,(4, 11)) or Isotachophoresis (ITP,
(14)) in a relatively *non-restrictive* medium may be used to provide a slightly
different type of *charge fractionation.*

One-Dimensional MM

Resolution:

Over the last 10 years, serum has been widely analyzed by PAGE using
conditions simular to those originally described by Davis (5); about 30 serum
proteins can be resolved in this way. Since in PAGE in multiphasic buffer
systems (MBS) (4, 11) the number of bands appearing with reliable R_f in the
Separation Gel is limited to the number of components which are *stacked* in
the Upper Gel (4, 11), it appeared of interest to determine the percentage of

serum protein stacked in Davis' buffer system. This was found to be perfectly adequate — approximately 90% — based on densitometry of stained protein in a Stacking Gel. The range of mobilities within which proteins are stacked was widened maximally, by decreasing the *lower stacking limit* (designated in (24) as RM (1, 4)) from 0.090 (system B) at pH 9.6 to 0.030 (system 2964.3) at pH 10.4. However, the percentage of serum proteins in the stack was not significantly increased (figure 1 of ref. (14)) by this maneuver, nor by reduction in molecular sieving by use of an agarose-polyacrylamide gel (14) in lieu of the customary 3.125 %T, 20 %C gel. Analytical PAGE at pH 10.2 (system B, 4—10 %T, 3 %C) was carried out on 2.5 μl serum/gel. A maximum of 22 R_f values (distinct bands) could be identified in 4 and 5 %T gels (table 2). PAGE at pH 11.0 (system 2964.3 (24), 4—7.5 %T, 3 %C) yielded 12 serum components on the basis of R_f at the same gel concentrations.

There are some problems of MM involving only 25—50 components; these may lend themselves to analysis by 1-D MM, particularly, if the band of interest is a relatively major one and if it is clearly separated from its neighbors. A representative example is the detectability of *de novo* synthesized proteins as a function of time after infection of HeLa cells with vaccinia virus (figure 1 of (29)). The simple 1-D method seems particularly useful when applied to gels in buffers containing sodium dodecyl sulfonate (SDS), since the disaggregation products formed by SDS usually cover a far narrower range of molecular sizes than the intact proteins of a multicomponent system. Reaction of proteins with SDS results in (near) equalization of mobilities among components, by means of equalization of molecular charge density, due to binding of SDS. After incubation of the protein mixture with SDS under proper conditions, the free electrophoretic mobilities of all components become nearly identical, and higher than the free mobilities of the native molecules. SDS also produces loss of helicity with unfolding of the specific protein conformations to produce, uniformly for all components, a *random coil* conformation (53). Since the *random coil* conformation implies an increase in molecular size compared with the globular conformation, diffusion coefficients for all components should decrease, implying narrower bands and improved resolution. The combination of the SDS method with the gel-slab technique (which improves precision of measurement of small differences in mobilities) has allowed for sufficient resolution on 1-D gels, e. g. in application to virus-infected bacterial systems (30, figure XX) involving approximately 25 components.

Pore Size:

What is the optimal gel concentration (effective *pore size*) for a complex mixture of macromolecules of varying sizes?

Methods for *size* fractionation using PAGE are based on the relationship between the fraction of the gel *available* to the molecule and the size of the molecule (Equ. 10 of ref. (15)). If the gel is too restrictive, then a large proportion of the sample is unable to enter the gel. This may result in excessive protein concentrations on the top of the gel, *clogging* of the pores, coprecipitation with other proteins, and finally, deformation of the gel (especially if the mechanically labile Stacking Gels (4, 11) are used). Deformation of the gel may result in separation of the gel from the walls of the apparatus, and complete loss of fractionation. This was found to be a severe problem for isoelectric focusing and isotachophoresis with high sample loads. Thus a moderately *restrictive* gel may be used (for serum, e. g. a 7 %T, 5 %C gel). Such gels of *medium* pore sizes do not exploit the *optimal pore size* (15, 31) for the separation of a relatively small number of component pairs. The number of detectable serum components obtained depends on the particular gel concentration used (table 2).

Table 2. Number of Serum Protein Components in Analytical PAGE (System B, 3 %C) based on

	A		B				
No.	K_R	Y_0	R_f				
			4	5	6	7.5	10 %T
1	0.406	11.37	0.017	0.020	0.022	0.010	0.010
2	0.254	3.43	0.051	0.037	0.036	0.020	0.024
3	0.192	2.23	0.068	0.055	0.065	0.043	0.034
4	0.163	1.96	0.085	0.088	0.090	0.060	0.043
5	0.112	1.38	0.119	0.110	0.108	0.086	0.053
6	0.079	1.10	0.170	0.148	0.140	0.108	0.068
7	0.09	1.29	0.262	0.185	0.160	0.124	0.097
8	0.079	1.47	0.329	0.245	0.220	0.144	0.108
9	0.070	1.46	0.393	0.300	0.290	0.160	0.121
10	0.071	2.00	0.440	0.365	0.320	0.175	0.145
11	0.093	3.28	0.487	0.407	0.361	0.211	0.165
12			0.526	0.449	0.410	0.235	0.178
13			0.542	0.510	0.433	0.282	0.213
14			0.600	0.534	0.490	0.325	0.233
15			0.624	0.565	0.508	0.375	0.252
16			0.705	0.608	0.535	0.380	0.288
17			0.743	0.640	0.575	0.433	0.331
18			0.788	0.658	0.605	0.450	0.386
19			0.813	0.702	0.690	0.545	
20			0.895	0.785	0.720	0.631	
21			0.946	0.833	0.932		
22			0.990	0.950			

Note: The K_R and Y_0 values do not derive from the R_f values shown on the same lines.

"Step-function" gels:

In order to accomodate a wide range of molecular sizes (from less than 1000 to greater than one million for serum proteins), one can utilize an *open* pore gel (low gel concentration) applied above a *small pore* gel of higher concentration; this arrangement was designated as a *step-function* or discontinuous pore gradient gel. (In theory, it is possible to obtain very interesting results if the small pore gel is placed above the open pore gel. However, in practice, the *clogging* effect precludes this arrangement.) It has been claimed (25, 26) that resolution of serum is vastly improved if fractionation in a *medium pore size* is replaced by a *step-function-gel* of up to 4 gel concentrations. This arrangement has the advantage that a multiple-layer-gel is prepared more easily, and probably more reproducibly, than a continuous pore gradient electrophoresis (P-G-E) gel. Step-function gels of a low number of *steps* have the disadvantage, compared to continuous pore gradient gels, that relatively fewer component pairs are provided with the specific gel concentration, at which each is optimally resolved (T_{opt}) (15, 31). However, in our laboratory a repeat of a step-function gel fractionation (system B, 4 and 10 %T, 3 %C) very similar to one previously reported (26) to give a 50% increase in the detected number of bands for serum revealed only 20 protein components on the basis of R_f.

Pore gradients:

Another approach to size fractionation is the use of pore gradients, i. e. gels with a continuously variable gel concentration. Usually, linear gradients are used rather than non-linear ones since they are more easily formed by use of simple gradient formers, and because they simplify prediction of the instantaneous velocity and position of any molecule in a linear pore gradient (7). Of course, non-linear gradients can be and have been used, often inadvertently. The pore gradient gel increases the range of molecular weights which can be fractionated simultaneously relative to constant %T gels and, to a lesser degree, step-function gels. This advantage is offset by the smaller distance between those bands which do enter the gel. P-G-E gels involve a number of problems: 1. Technical problems in the formation of reproducible gradients, e. g. convection and variable polymerization efficiency along the gel due to the rise of the heat of polymerization with %T; variable and relatively low polymerization efficiency along the gel (even at high %T) due to the need to retard the initiation of the polymerization reaction during gradient formation; difficulties in the control of temperature, exposure to light and oxygen tension in the polymerization mixture during formation of the gradient, etc. 2. Uncertainty regarding the exact %T at any point on the gel, due to non-linearity of the gel gradient and due to failure of the gradient maker, such that the %T at the ends of the gradient may not correspond to the intended values. 3. Alteration of the electrical field by the pore gradient. 4. Uneven rate of migration of the front, as readily seen when MBS's are used. 5. Severe problems of obtaining sufficiently reproducible gradients to permit comparison of patterns for several different multicomponent mixtures. 6. Use of sucrose gradients to stabilize the gel gradient during formation: The superimposition of a viscosity gradient alters the Ferguson-plot relationship.

On grounds of the T_{opt}-argument (see above) a pore gradient gel can be expected to be superior in resolving capacity to single pore or, at least in a quantitative sense, to step-function gels when applied to multicomponent systems. The degree to which this superiority of the P-G-E gel becomes effective and practical depends on the choice of pore gradient or step-function gel used, on the distribution of molecular weights of the sample and on reproducibility. We found, e. g. that a pore gradient of 3.5–12 %T gives no significant improvement of resolution of serum as compared to a 7.5 %T gel. In contrast, Kenrick and Margolis reported the results of a 3–26 % pore gradient (27) which shows nearly twice as many components.

Summary of 1-D MM:

The limitations imposed by consideration of the number of theoretical plate equivalents, N, are particularly severe for the one-dimensional *mapping* methods. Even if we could achieve 2000 theoretical plates on a routine basis, this would only enable us to fractionate about 25 components if these were evenly distributed along the gel. With staining techniques, where the relative insensitivity of the dye only reveals a small, central portion of the concentration profile (16), we might be able to detect about 50 components if we *relax* the definition of resolution for 2 closely adjacent bands compared with that used by Giddings (9) or Svensson (28).

Two-Dimensional MM

Theoretical plates:

The concept of *theoretical plates* is very difficult to apply to the 2-D fractionation methods. If we apply two techniques in series, with theoretical plates of N_1 and N_2 respectively, what is the *equivalent* number of theoretical plates for the entire procedure? This is very difficult to answer. Entropy can be used as a general criterion for separation: this approach should be applicable to 2-D separations (54). Applying the concept of N intuitively, it would appear that the best we can hope for is $N_{max} = N_1 \times N_2$, i. e. that the resolution of the combined procedure would vary with the product of the respective number of theoretical plates. However, it is much more likely, that the *resolution* of the 2-D procedure will be closer to the sum of the number of theoretical plates (thus, we have an additive, rather than a synergistic effect). Many of the methods used to date may result in even less than a fully additive effect in terms of number of theoretical plates.

Consider the situation where the second dimension of fractionation was performed by the same method and under exactly the same conditions as the first dimension. Then, the distance migrated by the molecule in the second dimension would be exactly the same as in the first (if we neglect problems such as re-stacking of the proteins before the second dimension of fractionation). Then, all of the proteins would be spread out along a straight line, the diagonal or $x_2 = x_1$. If the length of the first and second dimensional gels is the same (L), then the effective length of the electrophoretic path for the combined procedure would be $\sqrt{2} \times L = 1.4L$. Thus, the total number of theoretical plates (which is proportional to path length) would be increased by 40%. In other words, this case (identity of 1st and 2nd dimension) results in far less than an additive effect in terms of number of theoretical plates. Fractionation efficiency would be even worse than in the 1-D case since the total migration distance is 2L and bandspreading correspondingly large (unless we re-stack in the second dimension). Indeed, one could have done better, in the same length of time, simply to use a 1-D gel of twice the length: at least this would have resulted in a doubling of N. Also, that would avoid the problems of obtaining a perfectly uniform gel slab (which, in this case, would be more difficult than obtaining a single uniform gel cylinder). Thus, we can evaluate methods in terms of how well they spread the protein spots away from the diagonal, how really *two-dimensional* they are either in practice or in computer-simulated fractionations (31).

2-D fractionation of *size isomers:*

Raymond has suggested the use of PAGE in one dimension, followed by PAGE in a second dimension employing a gel of a different pore size (32). This has been designated as the *orthacryl* approach, and indeed represented the first use of 2-D MM using PAGE. In the example (figure 8 of (32)) Raymond chose a 4% gel in the first dimension, followed by an 8% gel in the second dimension. Let us see how a series of *charge isomers* and how a series of *size isomers* would behave in this system. Molecules differing only in charge (free mobility, Y_0), but of the same size (K_R), would migrate with the same ratio of mobilities in the two dimensions, i. e. $(M_1/M_2)_{T=8} = (M_1/M_2)_{T=4}$. Thus, a series of charge isomers would end up on a line through the origin (though not the diagonal, as discussed above). On the other hand, for molecules with the same free mobility (at 0 %T), but different size (K_R), one would expect that the R_f in a second dimension (gel concentration T_2)

would exactly equal the square of their respective R_f in the first dimension, since $T_2 = 2 T_1$. Thus, the pattern for one such series would be a parabola, extending through the origin. If there were several homologous series of proteins, each with a different free mobility, then one would obtain a family of parabolas. If the %T in the 2nd dimension were 3 times the concentration in the first dimension, then these arcs (for the series of size isomers) would follow an arc described by $(R_f)_{3T} = (R_f)_T^3$. In general, if the gel concentration in the second dimension were k times that in the first dimension, the arc would have the form $(R_f)_{kT} = (R_f)_T^k$. This follows directly from the Ferguson relationship (15, 20)

$$\log R_f = \log Y_0 - K_R\, T$$

and the assumption that Y_0 is unity for all members of the series (33).

2-D *size fractionation* of *size and charge isomers:*

Let us now look at the results obtained for serum — a mixture of *size* and *charge* isomers. It is seen (figure 7 of (27)), that the proteins are distributed along a parabolic arc, as would be expected for a series of (predominantly) *size isomers.* This is not too surprising. The molecular weight range of serum proteins is very large, covering about 2 orders of magnitude, while the range of free mobilities is relatively small. However, there are several components distributed away from the main parabola, representing *charge* isomers, presumably of (nearly) the same size. Most of the spots are concentrated in a region encompassing about 40% of the total area of the gel slab. Thus, we shall give this method a rating of 40% compared with the theoretical ideal, where all of the gel surface area is used. Predominantly diagonal patterns also result from combinations between PAGE and P-G-E (23) since the alignment of bands is usually only slightly different in the 2 methods, as seen from a comparison of serum patterns between PAGE (figure 8 of (4)) and P-G-E (figure 3 of (34)). When both pH and pore size are different for the 2 dimensions of PAGE, one can obtain a significant improvement in the utilization of the entire surface of the gel slab (figure 8 of (35)).

2-D *charge fractionation* — *size fractionation:*

Margolis and Kenrick have used (figure 1 of (23)) PAGE under *"non-sieving"* conditions as a *charge fractionation* tool followed by a concave pore gradient. Under the conditions used, the serum pattern remains essentially diagonal. Possibly the approach would have been more effective had the first dimension of PAGE be carried out at a pH closer to the isoelectric points of the components.

Utilizing the same principle but an experimentally simpler procedure, Maurer and Allen (36) obtained a remarkably efficient macromolecular map of serum by use of a cellulose acetate electrophoresis — step-function PAGE fractionation (figure 1 of (36)).

2-D IFPA-PAGE:

The most commonly applied charge fractionation in 2-D-MM is IFPA. IFPA can be used effectively as a charge fractionation tool only if %T is low enough to minimize molecular sieving effects, so that proteins reach their ultimate pI-positions on the gel in the relatively short time allotted. A representative 2-D IFPA-PAGE serum fractionation is that of Dale and Latner (figure 2 of (37)). Kenrick and Margolis (figure 1 of (38)) fractionated serum proteins by IFPA-P-G-E. In both of these applications of 2-D *charge vs size* maps, at least one-half of the available slab area still remains unoccupied.

Better resolution was obtained by Wrigley (figure 3 of (39)) who used a 2-D IFPA-starch gel electrophoreis method to fractionate a mixture of gliadin components. However, once again, Wrigley's patterns also predominantly follow a diagonal geometry. Probably the most extreme case of exploitation of the use of 2 independent fractionation methods for 2-D MM is the IFPA-SDS-PAGE gel introduced by Stegemann (40). The use of SDS in the second dimension has not only the advantage of converting PAGE to primarily a size fractionation method, but it also aids in solubilizing the bands which have been isoelectrically precipitated in IFPA.

The success of removing bands from the diagonal on 2-D maps can be measured by calculating the familiar correlation coefficient (r) for the "x_1" and "x_2" coordinates for each species. If all *spots* fall on the diagonal, then r = 1. In the ideal case, the correlation should be zero.

Numerical identification of components on 2-D maps:

Methods of MM lend themselves to different degrees to a quantitative, numerical description of the pattern and identification of its components. Without such quantitative identification of the coordinates for each component (band or *spot*) and the assignment of statistical confidence limits to the identifying parameter(s), unambiguous characterization of a pattern and its components is not possible, and comparison between data obtained at various times or in different laboratories is difficult if not impossible. Band quantification and the assignment of confidence limits to terms of electrophoretic mobility is readily available for 1-D methods, once the conditions of gel formation and of electrophoresis are standardized and controlled (1, 7). No such techniques of quantification have been applied to 2-D methods to date. Therefore, 2-D maps suffer from subjectivity of pattern interpretation. However, it should be possible to define spots on such a fingerprint quantitatively in terms of a coordinate system and to assign confidence limits to the position of each spot and to inter-spot distances etc. The entire system of relative distances and angles between components that constitutes the *pattern* could then be stored in a computer, the statistical variability of the pattern could be calculated, and pattern variations could be detected by the computer in an objective fashion, on the basis of predetermined criteria with regard to the *baseline* and to the tolerated variance for a given component. The potential diagnostic and scientific consequences of such an interpretation of 2-D patterns would be considerable, as has been originally pointed out by Ornstein (4) with regard to 1-D gels. In view of the wide confidence limits for R_f, K_R and Y_0 observed in quantitative *1-D* PAGE (16, 41—43) it can be expected, however, that the confidence limits obtained for spots on 2-D maps may be even much wider, particularly when pore gradients are used.

Transverse Pore Gradient:

Kenrick and Margolis (27) and Thorun and Mehl (44) have introduced a technique which might be called *transverse pore gradient electrophoresis.* Here, the sample is applied along the entire edge of a 2-D gel slab, such that it will migrate electrophoretically into a wide spectrum of pore sizes (e. g. 5—15 %T) simultaneously (figure 1 a of (44)). Although this method uses a 2-D slab and produces a 2-D pattern, this should be classified as a 1-D separation technique since the sample is fractionated in only 1 direction or dimension. Each component appears as a smooth, continuous arc rather than as a band or a spot. In effect, on a transverse pore gradient gel pattern we can examine the relationship between mobility (or R_f) and %T directly and simultaneously rather than in different gel tubes. This has the advantage that there is little if any ambiguity in following the mobility of a particular species as a function of gel concentration. In contrast, when a multicomponent sample is subjected to electrophoresis in several different gel tubes, with discrete gel concentrations (e. g. 5, 6, 7, 8, 9 %T) there may be considerable ambiguity and uncertainty as to which band in a 5 %T gel corresponds to which band in a 6 %T gel etc. The assignment of bands may become impossible even if the increment in %T is only 1 %, and if the Ferguson relationship is used (e. g. figure 1b of (33), table 2 of this report). However, the transverse pore gradient has several drawbacks and unresolved problems which make it difficult if not impossible, to construct Ferguson plots for *quantitative PAGE* from transverse gradient gels. Differential swelling properties of different portions of the gradient (during fixation and staining) complicate the problem of obtaining accurate x_1, x_2 coordinates on any point, needed for calculation of mobility and %T. Occasional problems also arise in determining the course of an *arc* of a protein at a point of intersection. Less than ½ of the gel slab surface is effectively utilized.

In view of these difficulties, and the fact that we must use a 2-D apparatus for an essentially 1-D fractionation, this method has limited appeal. However, it may be very useful, when used in conjunction with conventional gel tube methods, in the assignment of bands to Ferguson plots. Also when the technical problems are solved, (e. g. correlation of horizontal position with %T), then this method should facilitate the rapid construction of Ferguson plots. Also, this method can be used as a rapid screening method to find the %T for optimal resolution of any two separate species (see (31)).

2-D Perpendicular Pore Gradient:

In theory, there is an optimal gel concentration which will provide optimal separation and/or resolution between any two species (31). In order to achieve a high degree of resolution of a multicomponent system, it would be desirable to have each pair (or cluster) of species fractionated at the appropriate T_{opt}. This suggests that the first stage (or dimension) of fractionation be used to assign (or allocate) various species to the appropriate T_{opt}. Consider the case of a homologous oligomeric series of proteins, differing in size but not in charge. If this is fractionated in an open pore gel (e. g. 5%T) the smaller species migrate more rapidly, the larger more slowly. If this gel is then applied to the upper surface of a transverse pore gradient, then the smaller molecules can be fractionated (in the second dimension, perpendicular to the first) in a high %T (small pore) gel, while the larger molecules will be fractionated in a relatively open-pore gel. In this scheme, a mixture of *charge and size isomers* will be distributed nearly uniformly over the gel surface. It appears that this method may be further improved by use of a pore gradient for the 1st dimension of fractionation. Thus, (figure 1, top panel) we have fractionation from top to bottom) in a 3−25% gradient in the strip shown at the left. After a suitable time, such that the fastest components have reached the end of the gel, we apply the strip to the slab and change the direction of the electrical field by 90°, so that migration proceeds downward into the gel slab. If the gradient of the 1st dimension of fractionation and of the gel slab are the same, then each molecule remains in the same %T and pore size as it had reached in the 1st dimension of fractionation. This arrangement is not mandatory, but it is easily achieved in practice, and in principle, should be close to optimum. If one uses the *dead-stop* or *pore limit* hypothesis (to which we do *not* subscribe (7)), then one could regard the 1st dimension as permitting each molecule to find its *maximally restrictive* pore size, followed by fractionation in this tight pore size. Although this method has several ideal properties in theory, it has not been tested, as yet, due to problems of apparatus design and construction. We have attempted to combine this approach to fractionation with several other desirable properties of the apparatus including 1. all Pyrex glass construction, necessary for satisfactory

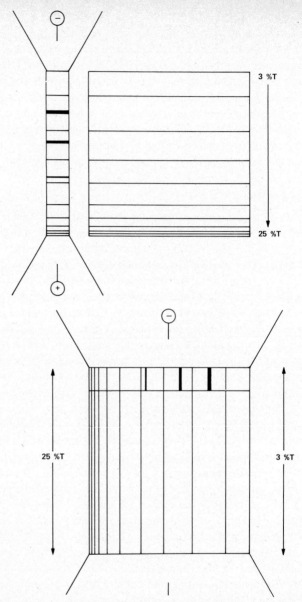

Fig. 1. Principle of two-dimensional perpendicular pore gradient electrophoresis. An identical pore gradient gel (e. g. 3–25 %T) is formed (simultaneously) in a rectangular cylinder (left top) and a slab (right top). P-G-E is carried out in the cylinder. The gel is removed from the tube and placed on the top of the gel slab (bottom). Cylinder and slab are oriented at a right angle to the orientation of the apparatus during polymerization and the 1st dimension of electrophoresis. Electrophoresis is then resumed in the 2nd dimension, as in the use of a transverse pore gradient. In theory, this should result in a nearly uniform distribution of the components of a multicomponent system over the slab.

wall adherence at acid and neutral pH, and at $0°C$, and for optimal temperature control, 2. provision for stacking in both 1st and 2nd dimension of
fractionation, 3. temperature control, 4. ability to rotate the direction of the
electrical field, without removing the gel from the apparatus, and involving a
minimum number of sealing and bonding steps. Remaining problems of this
method include 1. Difficulty of obtaining reproducible pore gradients, i. e.
problems of between-experiments reproducibility, 2. excessive complexity,
e. g. compared with use of cellulose acetate electrophoresis followed by P-G-E
in SDS-containing buffers.

2-D apparatus:

Present 2-D PAGE apparatus and techniques have not been developed to their
potentially *universal* applicability. They are deficient from the standpoint of
ease of operation. Further, most designs do not permit stacking in the 2nd
dimension of fractionation and thus fail to provide the narrowest possible
starting zones, with a resulting loss of resolution. Most apparatus, with
exception of the prototype device discussed above, are made of hydrophobic
plastics which do not provide adequate adherence of the gel to the wall over
the full useful spectrum of pH, temperature and gel concentration. Those few
apparatus models which are made of Pyrex glass require reassembly and
bonding of its components prior to each experiment. This approach, literally
using adhesive tape in some cases, should be replaced by designs that provide
the user with an assembled apparatus and the maximum number of prefabricated, self-sealing parts. The design should also provide for stacking the
components resolved in the first dimension before fractionation in the second.

Two-Stage MM

One approach to MM that appeared capable of overcoming the overlap of
distributions of major and minor components involves the use of preparative
PAGE, P-G-E, IFPA or ITPPA. Elution of component fractions from a
preparative first stage is followed by analytical scale fractionation, using again
any of the gel fractionation methods. The promise of this approach lies in the
fact that it allows for a 2nd dimensional analysis not only at various gel
concentrations but also at various loads. Analysis of fractions of the 1st
stage by use of variable aliquots ot the 2nd stage should allow one to locate
minor components by analysis at high concentration at the second stage,
using fractions which are relatively free of the major components.

Another advantage of this approach is that preparative PAGE apparatus is
relatively well developed and commercially available. In particular, a *micro-*

preparative device of 1.3 cm^2 gel surface and discontinuous elution (modified Fractophorator apparatus, "Apparatus A") (12)) seemed useful in this context. Isotachophoresis was carried out in a preparative device designed for continuous elution (Polyprep apparatus, Buchler Instr., "Apparatus C", using the procedure of Appendix III of (12)) since this design lends itself better to the mechanical instability of *"non-sieving"* gels.

Three approaches to 2-stage fractionations were tested: Preparative PAGE, P-G-E or ITPPA were the first stages; analytical PAGE at various pore sizes was then used to fractionate the eluates. Particular attention was paid to the increase in resolution that could be expected when the first stage of fractionation operated on a principle different from the second. This is the case when IFPA or ITP are used as the first dimension of fractionation, since both of these methods exploit primarily net-charge differences between macromolecules.

The 2-stage P-G-E (system B, 3.5—12 %T, 3 %C) — PAGE analysis of serum on a modified Fractophorator apparatus (figure 2) failed to show substantial improvement of resolution compared to 1-D analysis. This analysis is dependent on the choice of %T for the 2nd stage fractionation of various eluate fractions. Different eluates contain (mainly) species of differing R_f. Several suitable %T were selected for each eluate so that Ferguson plots for each component could be constructed. Usually, the eluate fractions contained 4—5 components. Thus 4—5 bands in each of 4—5 gels of different %T were obtained upon 2nd stage analytical PAGE. The assignment of bands to components was not possible without an element of guesswork in spite of the rules of assignment (see Methods, Section 11) followed. The lines in the left and center plots of figure 2 represent a very conservative assignment and thus the minimum number of components. About 70 K_R-Y_0 pairs (upper right) resulted from computation of the Ferguson plots. However, when the confidence regions, defined by an ellipse with semi-axes equal to the σ_{K_R} and σ_{Y_0} are shown, the *dots* become fuzzy spots. If we require non-overlap of these ellipses, to be confident that 2 bands are different, then only 22 serum components could be distinguished. This is an exceedingly difficult problem from the point of view of statistical analysis, in view of the large number of pairwise comparisons. It is possible that we can increase the sensitivity by consideration of the joint 95% C. L. for K_R and Y_0 (including the correlation of errors in the 2 parameters), by suitable pooling of data, and by consideration of some of the correlation of errors *within experiments.* Nevertheless, the data do not indicate that P-G-E-PAGE promises an increase in resolution which would would be commensurate to the enormous labor required for the 2nd stage of such fractionations: 100—400 fractions of the 1st stage in a representative case are distributed over the same total number of analytical gels at each of 4 gel concentrations; with 3—5 bands per analytical gel this requires the measurement of 300–5000 band distances and calculation of 300–5000 R_f values. Further, one must evaluate which bands in each of the 4 gel concentrations belong to a single component, compute the Ferguson plots for each component and evaluate identity or non-identity for each of the resulting K_R-Y_0 sets.

The results of 2-D PAGE-PAGE fractionation of serum, using at the 1st stage MBS B (20), 7.5 %T, 3 %C, were very similar to those of the P-G-E-PAGE fractionation depicted in figure 2. The number of components, distinguishable on the basis of K_R and Y_0 by the above criteria is no larger than that obtained from a 1-D fractionation, and certainly not sufficient to justify the labor involved in such analyses.

ITPPA-PAGE appeared to possess particular promise not only since the two stages of fractionation differed in principle of operation, but also since ITPPA has considerable load capacity (200 mg/1.3 sq cm and 300 mg/10 cm^2 were used on Apparatus A and C respectively). ITPPA-PAGE indeed revealed more components of serum than P-G-E-PAGE. For example, at 6 %T, the ratio in a representative experiment was 25/9 (figure 9 of (14)). This method revealed K_R-Y_0 sets (and presumably serum components) which were not seen by the other methods (table 1 of (14)). This may be an artifact due to Ampholine binding to protein. But even if this is not the case, 75 % of the K_R-Y_0 pairs were not significantly different by the above criteria.

Table 3 summarizes the results of 2-stage fractionations of serum. The preliminary nature of this examination of 2-stage MM does not allow us to rule out the possibilities that a wider range gradient for the first stage (27),

Table 3. Number of Serum Components Detected by 2-Stage Techniques.

Stage		No. of bands	No. of components based on K_R-Y_0
First	Second		
PAGE (pH 10.2) (pH 11.0)		22 12	11 12
(step-function)		20	
P-G-E		28	
ITPPA		14	
PAGE	PAGE		75 (30)
P-G-E	PAGE		80 (25)
ITPPA	PAGE		20

Note: numbers in paratheses designate the number of components such that there is no overlap of ellipses with semi-axes = 1 SD of K_R and Y_0.

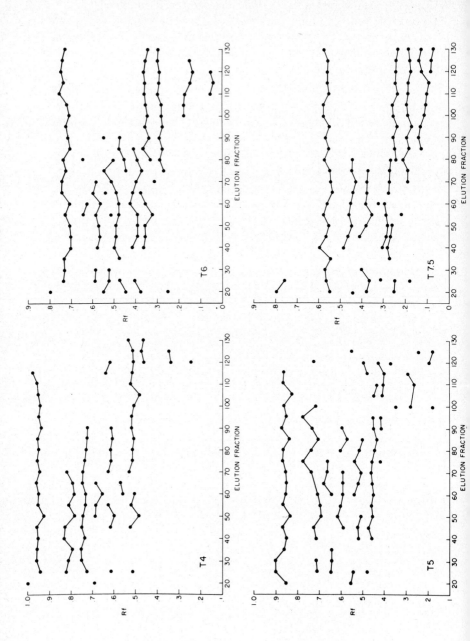

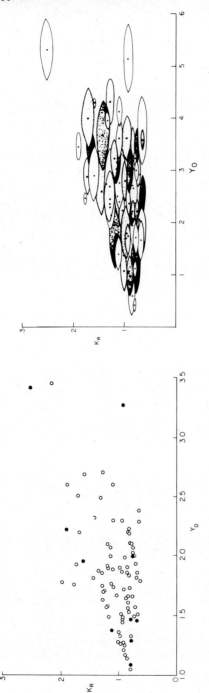

Fig. 2. Two-stage P-G-E (system B, 3.5—12 % T, 3 % C)-PAGE fractionation of serum proteins.
Load: 35 mg serum protein/1.3 cm² of gel.
Current: 4.6 ma/cm².
Fractions were collected at 10 min intervals.
Left and middle: R_f of all bands appearing in PAGE fractionation of the eluate fractions in 4—7.5% T, 3% C gels.
Upper right: Plot of K_R vs Y_0 for each of the eluate components.
Lower right: Joint confidence envelopes (± 1 S. D.) of K_R and Y_0 for each of the the eluate components.

or use of different load levels at the 2nd stage might provide improved resolution. However, the results are in no way promising, especially in view of the tedious and expensive nature of this approach. Using a relatively small number of preparative fractions (100 or less) we found 3–5 components in most eluate fractions even without an increase in load of the analytical gels at the second stage. This number was too high to allow for an unequivocal assignment of R_f values to components when adjacent eluate fractions were analyzed at different gel concentrations. A larger number of eluate fractions and minimization of band width at the time of elution at the 1st stage may be required to obtain fewer components per fraction. Also, a change of buffer pH between the 1st and the 2nd stage should improve resolution.

The problem of assignment of R_f values at different %T to various proteins does not arise in the transverse P-G-E slab technique. Accordingly, this method might be used for the second stage of fractionation — if a transverse P-G-E method can be developed that is both convenient and reproducible, for application to large numbers of routine fractionations (e. g. commercially prefabricated transverse P-G-E slabs).

Conclusions

The resolution of multicomponent systems containing more than 25–30 components, such as serum, remains unsolved. Available methods of poly-acrylamide gel electrophoresis have been evaluated from the standpoint of 1. resolving power, measured (at least in 1-D MM) as the the number of theoretical plates, 2. the mutual overlap of band distributions for the major and minor components, 3. quantitative identification of coordinates of components and of patterns, 4. optimization of gel concentration for the resolution of each pair of components, 5. the advantage of coupling a *charge fractionation* with a *size fractionation* and 6. practicability and ease of operation.

Among present methods, IFPA or cellulose acetate electrophoresis in the first dimension, followed by PAGE or P-G-E in SDS buffers in the second dimension appear to satisfy criteria 1., 5. and 6. and provide the best overall performance. The 2-D perpendicular gradient gel seems promising from the viewpoint of criteria 1. and 4. Criterion 3. has not yet been fulfilled by any of the 2-D methods, but may potentially be fulfilled by all of them.

Two-stage fractionation, employing a charge fractionation (e. g. ITPPA) as the 1st step and transverse P-G-E gels as the second, can potentially meet the first 5 of those criteria, but this approach appears excessively laborious and thus fails by the 6th criterion.

References

(1)   Chrambach, A. and D. Rodbard, Science *172*, 440 (1971).
(2)   Finlayson, G. R. and A. Chrambach, Anal. Biochem. *40*, 196 (1971).
(3)   Doerr, P. and A. Chrambach, Anal. Biochem. *42*, 96 (1971).
(4)   Ornstein, L., Ann. N. Y. Acad. Sci. *121*, 321 (1964).
(5)   Davis, B. J., Ann. N. Y. Acad. Sci. *121*, 404 (1964).
(6)   Schultze, H. E. and J. F. Heremans, Molecular Biology of Human Proteins, Am. Elsevier Publ., New York, Vol. I (1966), Vol. II (1972).
(7)   Rodbard, D., G. Kapadia and A. Chrambach, Anal. Biochem. *40*, 135 (1971).
(8)   Smithies, O., Biochem. J. *61*, 629 (1955).
(9)   Giddings, J. C., Sep. Sci. *4*, 181 (1969).
(10) Hedrick, J. L. and A. J. Smith, Arch. Biochem. Biophys. *126*, 154 (1968).
(11) Jovin, T. M., Biochemistry *12*, 871, 879, 890 (1973); T. M. Jovin, M. L. Dante and A. Chrambach, Multiphasic Buffer Systems Output, PB 196092, National Technical Information Service, Springield, Va., 1970.
(12) Kapadia, G. and A. Chrambach, Anal. Biochem. *48*, 90 (1972).
(13) Jovin, T. M., Annals N. Y. Acad. Sci., *209,* 477 (1973).
(14) Chrambach, A., G. Kapadia and M. Cantz, Sep. Sci. *7*, 785 (1972).
(15) Rodbard, D. and A. Chrambach, Proc. Nat. Acad. Sci. U. S. *65*, 970 (1970).
(16) Rodbard, D. and A. Chrambach, Chapter 1.
(17) Lunney, J., A. Chrambach and D. Rodbard, Anal. Biochem. *40*, 158 (1971).
(18) Catsimpoolas, N., Annals N. Y. Acad. Sci., *209,* 65 (1973).
(19) Hartman, B. K. and S. Udenfriend, Anal. Biochem. *30*, 391 (1969).
(20) Chrambach, A. and D. Rodbard, Anal. Biochem. *40*, 95 (1971).
(21) Abramson, H. A., L. S. Moyer and M. H. Gorin, Electrophoresis of proteins and the Chemistry of Cell Surfaces, Hafner, New York 1942.
(22) Rodbard, D., C. Levitov and A. Chrambach, Sep. Sci. *7*, 705 (1972).
(23) Margolis, J. and K. Kenrick, Nature *221*, 1056 (1969).
(24) Jovin, T. M., M. L. Dante and A. Chrambach, Multiphasic Buffer System Output, National Technical Information Service, Springfield, Va., PB No. 196085 to 196091, and PB No. 203016, 1970.
(25) Wright, G. L., Jr., K. B. Farrell and D. B. Roberts, Clin. Chim. Acta *32*, 285 (1971).
(26) Wright, G. and W. Mallmann, Proc. Soc. Expt. Biol. & Med. *123*, 22 (1966).
(27) Margolis, J. and K. G. Kenrick, Anal. Biochem. *25*, 347 (1968).
(28) Svensson, H., J. Chromatogr. *25*, 266 (1966).
(29) Moss, B. and N. P. Salzman, J. Virol. *2*, 1016 (1968).
(30) Laemmli, U. K. and M. Favre, J. Mol. Biol., *80*, 575 (1973).
(31) Rodbard, D., A. Chrambach and G. H. Weiss, Chapter 2, 3., this volume, p. 62.
(32) Raymond, S., Ann. N. Y. Acad. Sci. *121*, 350 (1964).
(33) Ugel, A., A. Chrambach and D. Rodbard, Anal. Biochem. *43*, 410 (1971).
(34) Margolis, J. and K. G. Kenrick, Nature *214*, 1334 (1967).
(35) Kaltschmidt, E. and H. G. Wittmann, Anal. Biochem. *36*, 401 (1972).
(36) Maurer, H. R. and R. C. Allen, Clin. Chim. Acta *40*, 359 (1972).
(37) Dale, G. and A. L. Latner, Clin. Chim. Acta *24*, 61 (1969).
(38) Kenrick, K. G. and J. Margolis, Anal. Biochem. *33*, 204 (1970).
(39) Wrigley, C. W., Biochem. Genetics *4*, 509 (1970).
(40) Stegemann, H., Z. Analyt. Chem. *350*, 917 (1969).
(41) Corvol, P., A. Chrambach, D. Rodbard and W. Bardin, J. Biol. Chem. *246*, 3435 (1971).

(42)   McIlwaine, I., D. Rodbard and A. Chrambach, Anal. Biochem., *55*, 521 (1973).
(43)   Chrambach, A., R. A. Yadley, M. Ben-David and D. Rodbard, Endocrinology, *93*, 848 (1973).
(44)   Thorun, W. and E. Mehl, Biochem. Biophys. Acta *160*, 132 (1968).
(45)   Rony, P. R., Sep. Sci. *5*, 121 (1968).
(46)   Boyde, T. R. C., Sep. Sci. *6*, 771 (1971).
(47)   Chrambach, A., G. Kapadia, J. Pickett, M. L. Schlam and N. A. Holtzman, Sep. Sci. *7*, 773 (1972).
(48)   Chrambach, A., R. A. Reisfeld, M. Wyckoff and J. Zaccari, Anal. Biochem. *20*, 150 (1967).
(49)   Awdeh, Z. L., Sci. Tools *16*, 42 (1968).
(50)   Myerowitz, R. L., A. Chrambach, D. Rodbard and J. B. Robbins, Anal. Biochem. *48*, 394 (1972).
(51)   Giddings, J. C., Anal. Chem. *39*, 1027 (1967).
(52)   Determann, H. and A. Walch, Z. Naturforsch. *27b*, 683 (1972).
(53)   Reynolds, A. and C. Tanford, J. Biol. Chem. *245*, 5161 (1970).
(54)   de Clerk, K. and C. E. Cloete, Sep. Sci. *6*, 627 (1971).

Appendix I: Methods

1. PAGE apparatus: Analytical scale PAGE and isotachophoresis on PA gel (ITPPA) were carried out in the *Polyanalyst* apparatus (Buchler Instruments Inc., Fort Lee, N. J.) as described (20). Analytical P-G-E was carried out in the partitioned slab apparatus described elsewhere (7, 47). Electrophoresis on a μg-preparative scale was carried out in a redesigned (12) "Fractophorator" (Buchler Instruments) apparatus (designated as Apparatus A); or in a *Prep-Disc* apparatus of 1.7 cm^2 gel surface area (Canal Industrial Corp., Rockville, Md., Apparatus E). Mg-preparative PAGE employed a *Polyprep 100* (Buchler Instruments) apparatus (designated as Apparatus C in Appendix III to (12)). Preparative ITPPA was conducted in preparative Apparatus A and C.

2. Buffer systems: Multiphasic buffer system B (20) or 2964.3 (24, 14) were used. System B is a Tris-system (5), modified for operation at 0°C, the properties of which were determined by the computer program of T. M. Jovin (11). System 2964.3, generated by this program, is operative at a stacking pH of 10.4, and at a separation pH of 11.0, 0°C (24).

3. PAGE procedures: Polymerization and electrophoresis were carried out as described previously (20). Gel concentrations in the Resolving Gels ranged from 4 to 7.5 %T, 3 %C; Stacking Gels were 3.125 %T, 20 %C (5). "Step-function-gels" of 10 %T and 4.75 %T, 3 %C, were polymerized on top of each other (25). Fixation and staining employed either 0.05% Coomassie Blue in 12.5% TCA (48, Footnote 4 of (20)) or 0.1% Amidoblack in 7.5% acetic acid (5, 33). Densitometry of stained gels was carried out as described (17). Integration of densitometric patterns was performed by weighing of the recording paper under the peaks. The characteristic R_f values for each band

were measured, and the retardation coefficients (K_R) and y-intercepts (Y_0) were computed as described previously (20). Joint confidence envelopes for K_R and Y_0 were constructed using ± 1 standard deviation (S. D.) (note: the correlation of errors in K_R and Y_0 (16) was ignored).

4. Analytical P-G-E in the partitioned slab apparatus: The procedure of P-G-E followed that described previously (7, 47). System B (20) and a gel concentration gradient between 3.5 and 12 %T, 3 %C was used.

5. Determination of the degree to which components of a multicomponent system are stacked: Serum was subjected to PAGE in Stacking Gels of system B (operative pH 9.5) and 2964.3 (operative pH 10.4); gels were stained and subjected to densitometry. The area under the peak representing the stack was integrated and compared with the total density of the gel. Percent of protein in the stack was estimated as the ratio of peak area to total area.

6. Procedure of analytical ITPPA: ITPPA was conducted as described (14) in buffer systems B and 2964.3. Gels of 3.5 %T, 5 %C or the 0.25 % agarose-2%-polyacrylamide copolymer (0.2% Bis) (14) were used. Ampholine (pI 6−8) was admixed to the sample of analytical scale gels in amounts varying from 40 to 100 μl. Gels were stained with bromphenolblue by the procedure of Awdeh (49) or sliced and subjected to pH analysis (2, 3).

7. Procedure of ITPPA in the Preparative Apparatus A: The procedure of PAGE in the preparative apparatus A used was that described elsewhere (12).

8. Preparative PAGE in the Prep-Disc apparatus: The procedure of polymerization and electrophoresis followed the principles of operation previously described ((12) Appendix II). System B (20), a 6 ml Resolving Gel and 2 ml Stacking Gel were used. Load: 35 mg serum protein. Electrophoresis was conducted at 6 ma. Elution Buffer flow rate was 0.235 ml/min. Fractions were collected at 10 min intervals.

9. Preparative ITPPA in the Polyprep-100 apparatus: The procedure of PAGE in the Polyprep-100 apparatus used was that previously described ((12) Appendix III).

10. P-G-E in the Preparative PAGE Apparatus A: The general procedures of pore gradient formation (7, 47) and of PAGE (20) were followed. Equal volumes (1.7 ml) of polymerization mixtures of 3.5 and 12 %T, 3 %C, were used to form a linear pore gradient gel. The TEMED concentrations were 40 and 15 μl/100 ml gel in the 3.5 and 12 %T gels respectively. The sample load was 35 mg serum protein, 500 μl of 50% sucrose. Electrophoresis was conducted at 6 ma/cm^2. Fraction Time: 10 min. Elution buffer I was a 1:4 dilution of Resolving (Lower) Gel Buffer; Elution Buffer II contained 168.3 g Tris, 180 ml 1$\underline{N}$ HCl and 250 g sucrose/liter (pH (25°C) = 8.11).

12. Calculations: Measurements of R_f and computations of K_R and Y_0 were carried out as described previously (20). The assignment of bands to Ferguson plots was made as follows: A) Each eluate fraction was analyzed on several gels with %T values differing by 1% increments, to minimize the possibility that the relative positions of bands in the pattern would change abruptly from one gel to the next. B) The positions of bands within the pattern, and the band thickness, intensity and color shades were considered. C) All R_f values obtained from the various eluate fractions were plotted on a semi-logarithmic scale vs. %T and combined by lines wherever possible. These lines were then tested and selected on the basis of the above criteria.

Chapter 4. Isoelectric Focusing and Isotacho-phoresis in Polyacrylamide Gel

4.1 Isoelectric Focusing in Flat Beds of Poly-acrylamide Gels

Olof Vesterberg

Most applications of isoelectric focusing of proteins have been made in columns stabilized with density gradients (IFDG). (For review articles see references 1, 2, 3.) However, since 1968 an increasing number of papers has been published where polycrylamide gel is used for stabilization. There are several reasons which have promoted the development of isoelectric focusing in polyacrylamide gel (IFPAG or PAGIF). A comparison is here made between isoelectric focusing in gels and in density gradient columns.

1. *Quick focusing.* Less than one hour is necessary for focusing with (IFPAG) (15). However, it should be kept in mind that the time necessary, and the resolution are influenced by the distance between the electrodes or electrode compartments. This has been discussed recently by Vesterberg (4). It can be mentioned that focusing in density gradients can also be obtained within one hour by using a short distance between the electrodes (5). 2. *High resolution.* Especially when using the high voltage procedure (IFPAG) (i. e. more than 800 Volts) very narrow bands are obtained (15). The resolution that can be measured or fixed, during focusing or directely thereafter, is most important for the experimentor. With the density gradient technique it is possible to record the focusing even when the current is on (5, 6, 7). However, this requires expensive instrumentation. On the other hand, a very high degree of resolution can be retained by simple means by fixing and staining of the proteins directely after IFPAG. 3. *Simple and inexpensive instrumentation* can be used for IFPAG. Many have used the same instruments as they already have for disc electrophoresis. However, this means focusing in gel rods. According to this and earlier comparisons with focusing in flat bed or thin layer of polyacrylamide gel most arguments are in favour of the latter technique (8). Because this technique differs quite considerably from the gel rod procedure it is not possible to make judgements of IFPAG without experience of the flat bed procedure. 4. *Less expensive* for analytical purposes as far the costs for chemicals are concerned. 5. *Many samples* can be handled and compared in a short time, especially when using the flat bed procedure. 6. The *staining procedures* used in connection with the gel techniques allow

detection of μg-quantities of protein or even less (8). 7. The gel techniques can be *combined with zymogram procedures* for specific detection of certain biological activities (9, 10). 8. The gel techniques can be *combined with immunological procedures* for detection of proteins (11, 12). 9. The gel techniques can be *combined with electrophoresis in another gel at right angle* (13), and if desirable in a gel gradient (11) to increase the resolution.

The capabilities of the IFPAG have not been fully explored due to the fact that no ideal instrument is yet commercially available. Some important improvements of the technique can be expected in the near future*.

Materials and Methods

In principle the same procedures as described recently by Vesterberg (8) were followed. The following changes can be notified.

Preparation of gel and apparatus (8). Sucrose was added to get a final concentration of 12% (w/v) in the gel. The gels were made 2 mm thick by using a gasket of this thickness instead of 1.5 mm. The gel glass plates were of the size $25 \times 12.5 \times 0.1$ cm. An extra glass plate 0.3 cm thick was used as a support for the gel glass plate at polymerisation to get a more even and solid support resulting in gels of more even thickness. The cooling plate was made of all glass and of a larger size to fit the new gel glass plates. Care must be taken to prevent leakage of the cooler.

Proteins. Samples of cerebrospinal fluid (CSF) were obtained in the Neurologic Clinic at Karolinska sjukhuset, Stockholm. These were concentrated about 8 times and aliquots of 0.03 ml were used for focusing. Urine from apparently healthy persons and from persons with various renal disorders, mainly tubular damage, were dialysed overnight and concentrated about 25 times. Aliquots of 0.1 ml were used for focusing as will be described (Vesterberg and Nise). Albumin and ovalbumin were from the sources earlier mentioned (8).

Application of sample. a) in pieces of porous paper (up to 0.15 ml) as earlier described (8) b) on the surface as a spot or a long streak (up to 0.1 ml) at right angle to the electrodes. c) in basins up to 0.25 ml (15).

Isoelectric focusing

In order to check the time necessary for focusing some protein samples on every gel were applied in two positions on one track close to each electrode. Usually 2 spots each of 2 μl of a hemoglobin solution containing roughly

* When this manuscript was in press a suitable instrument called Multiphore was presented by LKB-Produkter.

7 mg/ml of hemoglobin are used for this purpose. The joining of the hemoglobin zones can be followed during focusing. The experiment is then allowed to continue for another 15 minutes to give also slower migrating proteins a chance to focuse.

The maximum voltage that can be utilized is a question of how many watts that can be tolerated by the system. Since $W = V \cdot I$, where W = vatt and V = volt and I = current in ampere the following factors should be considered. a) *efficiency of heat transfer per cm² of gel surface* from the gel to the cooler; b) *surface area of the gel;* c) *thickness of the gel;* d) *pH range and concentration of Ampholine;* e) *distribution of the field strength between the electrodes.*

Usually focusing was made by circulating cooling water at $+ 10^0$ C through the cooling plate. The voltage was increased step-wise so that the heat was never allowed to exceed 50 W. (figure 1). When a distance of 10 cm was used

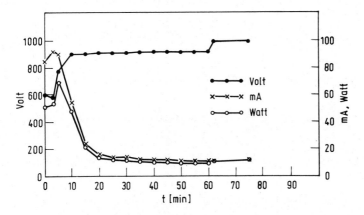

Fig. 1. Record of Volt, mAmpere and Watt during focusing in polyacrylamide gel for the pH range 3–10. A mixture of Ampholine pH 3–10, 2.8 ml; pH 8–10, 0.4 ml; pH 5–7, 0.2 ml; pH 4–6, 0.2; was used to get an average concentration of 2% (w/v). The gel measured 24.5 × 11.2 × 0.2 cm. A distance of 10 cm was used between anode and cathode. Hemoglobin applied at the electrodes joined after 45 minutes.

between the electrodes about 500 Volt was used at the start and 1000 Volt was used as a final voltage.

Measurement of pH. In principle the same procedures as described earlier were used (8). Care was taken to measure at 10^0 C both with the surface electrode and on the eluted samples.

* The carrier ampholytes used were Ampholine (LKB-Produkter 161 25 Bromma, Sweden).

Staining. Coomassie blue R 250 was used as a protein stain (8). For some proteins, e. g. from urine, this method gives a higher sensitivity than other methods tried (15, 16).

Densitometry on stained gels was made with a Zeiss densitometer, PMQ-2 Program Chromatogram Spectralphotomer.

Results

After focusing of cerebrospinal fluid samples about 30 protein zones of varying stain intensity were visible. A densitometric recording is shown in figure 2. Varying patterns were obtained from different patients. Work is in

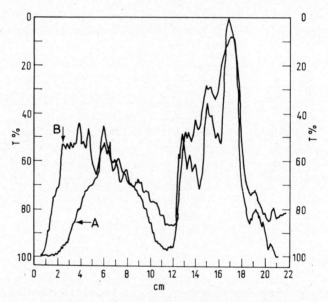

Fig. 2. Densitometric evaluation of CSF protein zones after isoelectric focusing at the same conditions described in figure 1 and after staining. The CSF samples were from A) a patient with trochlearis paresis, B) a patient with multiple sclerosis. Cathode to the left. Note the significantly higher amount of alkaline γ-globulins with pI > 7.5.

progress to evaluate the clinical significance of the various patterns obtained (17). All patients with the diagnosis Multiple Sclerosis showed a relative increase of alkaline γ-globulins (i. e. those with pI > 7.5 at 10⁰(C). This is also in accordance with a report by Delmotte (18).

Focusing of urinary samples revealed a marked variation in the number of protein zones detectable, as was also to be expected. Healthy persons usually excrete only minute quantities of proteins in urine. Therefore, with a given degree of concentration, only some of the major protein components can be detected. Because the significance of the individual proteins in urine of "healthy" persons is at present of limited interest, no effort was made to bring the proteins in the different samples to a comparable protein concentration, nor was adjustment made for differences in creatinine concentration. The aim of the present investigation was in the first instance to differentiate between glomerular and tubular patterns. The IFPAG method was found very suitable for this purpose as has also been found earlier (19, 20). The glomerular patterns being characterized by an increase in albumin, of course, and the tubular patterns by an increase in β_2-microglobulin. As this latter protein separated well from the other protein zones it is possible to estimate the degree of tubular damage by refering to the quantity of this protein. In advanced renal damage more than 40 protein zones were separable. Figure 3. (Vesterberg and Nise, Clin. Chem. in press Oct. issue 1973.)

The major hemoglobin components can be separated in accordance with an earlier report by Drysdale and Righette (21). It is possible by this technique to detect, e.g. sickle cell anemia which could be of value in clinical chemistry.

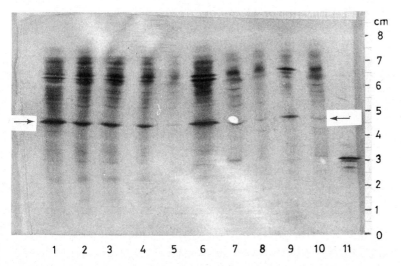

Fig. 3. The result of isoelectric focusing of urinary proteins. Before focusing samples of urine from different persons were dialyzed, and concentrated about 25 times. Samples of 0.1 ml were soaked into pieces of porous paper and applied on the surface close to the cathode. The conditoins for the gel and the focusing procedure were the same as described in the legend to figure 1. Cathode on top. Note the band marked with an arrow, that has been identified as β_2-microglobulin, which occurres at different concentration in the samples. Increased amount of this protein found in urine, is correlated to the degree of tubular malfunction of the kidneys.

Discussion

The advantages of the flat bed procedure of IFPAG as compared to the gel rod procedure are outlined in the "Concluding remarks". Here some comments on the preparation of the gels will first be made. Should the Ampholine be added before or after polymerization of the gel? Practical experience has shown that there seems to be little disadvantage of adding Ampholine before polymerization, especially when the high voltage procedure is used. However, polymerization at alkaline pH is not very easy when riboflavin is used as catalyst. Therefore, gels having an average pH above 7.5 ahould either be made by using ammonium persulfate or by polymerization of the gel before addition of Ampholine. The Ampholine can be added by soaking into the gel after polymerization.

Ampholine concentration: Too low a concentration may cause problems, e.g. "the plateau phenomenon" as has been shown by Chrambach (22) and by Drysdale and Righetti (23). When working in the pH range 3.5–9.5 it has been found suitable to use at least 2% (w/v) of Ampholine as a final concentration in the gel. Furthermore, to counteract "the plateau phenomenon" it is helpful to add a noncharged solute, such as sucrose, sorbitol or glycerol, to a final concentration of at least 10% (w/v) as has also been recommended earlier (8, 22). In order to get a more optimal distribution of the field strength between the electrodes is has been found advantageons to combine Ampholines of pH 5–7, 4–6 and 9–11 to get a pH gradient for the pH range 3–10. Each one should be added in an amount of 10% of the total amount of Ampholine to be used. Work is in progress to determine optimal compositions also for other pH ranges*.

A very important question is which electrolytes and which concentrations should be used at the electrodes? It is easier to give recommendations for the flat bed procedure than for the gel rod techniques. This is partly explainable by the fact that the volumes of the electrolytes used at the electrodes are usually smaller in the flat bed procedure. When working with the latter technique practical experience has shown that 1M H_3PO_4 at the anode and 1M NaOH at the cathode are suitable when a low and a high pH extreme of the pH gradient is to be expected. When less extreme pH values are expected in the pH gradient a weaker electrolyte, e.g. acetic acid or propionic acid in a concentration on 1M for more acidic, and 0.1 M for less acidic pH ranges e.g. pH 6–8, can be used at the anode*.

These recommendations are based on the premise that no zone with a low concentration of conducting solute should have a chance to develop between

---

* More detailed recommendations can be found in an article by O. Vesterberg in Science Tools, to be published end of 1973.

the electrodes. This is because such zones aquire a high field strength due to Ohm's law, which means that part of the potential is lost in a wrong place, and also leads to the risk of local overheating.

How about risk for oxidation of proteins? Gels made with riboflavin and especially with ammoniumpersulfate have oxidation capacity (24, 25). One way to circumvent this problem is to make the gel first, and then to allow the agents to diffuse out of the gel into water. Thereafter Ampholine can be soaked into the gel. Another way is to add reducing agents such as dithiothreitol (8). Control experiments to eliminate the influence of the catalysts on the proteins should be carried out in every case where such phenomena cannot be excluded.

It seems as if larger molecules can be focused in the flad bed gel than in the gel rod procedure. The explanation to this is that there is a possibility for such molecules to be transported in the upper surface layer of the gel, which most probably does not have as perfect a network structure as the interior of the gel (26).

However, sometimes there is risk for formation of aggregates or of large size multimers, which is often especially pronounced in certain pH ranges. In such cases various points of application should be tried. Other remedies are reducing agents and solubilizing agents, such as ethylene glycol (27) and other ones, e.g. urea and nonionic detergents (28). The acrylamide gel appears also to have some solubilization effect. Very often quite high protein concentrations are obtained locally in the zones without presence of protein precipitation. For example let us assume a protein zone with the dimensions 0.1 X 0.2 X 0.5 cm which gives a volume of 0.01 ml. Such a zone can very well contain 0.4 mg of protein, which makes a final concentration of more than 4% at the maximum, because the protein concentration can be assumed to have a distribution of a Gaussian curve.

Due to the high resolving power of the isoelectric focusing method, and especially of the high voltage IFPAG (15), more than one protein zone is often detected even in so called "pure" protein preparations. Many different forms of multiple molecular forms of proteins have been discovered. A comprehensive summary of the origin of such forms was recently presented (4). It can be recalled that by isoelectric focusing molecules differing by only a fraction of a net charge at a given pH can be separated (29). As partial denaturation or ageing of proteins is often supposed to start with changes in charge, e.g. deamidation, isoelectric focusing can be expected to be a sensitive tool for checking the native state of protein preparations.

The IFPAG method is becoming valuable in genetic and taxonomic studies, e.g. for plant proteins and in microbiology (30, 31, 32).

The high resolving power of the technique also seems promising for the use of the method for specific diagnostic purposes in clinical chemistry (33). Here we are just at the very beginning.

Concluding remarks on the comparison of the gel rod and flat bed procedure for IFPAG

1. If desirable carrier ampholytes can be added to the gel after polymerization by soaking.
2. The electrolytes used at the electrodes and the concentration thereof are not very critical.
3. No risk for the proteins to come in contact with the extreme pH of the electrode vessels at application of protein.
4. The proteins can be applied in many ways and at different points between the electrodes. Perhaps the most evident advantage of this is that proteins can be applied at "both ends", i.e. close to the anode and the cathode. When such proteins have joined the experiment can be regarded as finished.
5. All samples are focused under identical conditions and, as they can be run side by side, comparison is simple.
6. The time-consuming handling of many gel rods is avoided.
7. The flat bed is easier to handle for pH measurement, staining, destaining, and sectioning.
8. The gels are less likely to get scratched or broken, e.g. during staining.
9. The flat bed gel can retain its original size even after destaining.
10. The result with the flat bed gel may be conveniently recorded by photography and by densitometric evaluation.

References

(1)   Haglund, H., *In "Methods of Biochemical Analysis"*. Ed.: D. Glick, Vol. 19, Wiley (Interscience), New York, N.Y., 1971, pp. 1–104.
(2)   Vesterberg, O., *In "Methods in Enzymology"*, Ed.: W. B. Jakoby, Vol. 22, Acad. Press. New York, N.Y. 1971, pp. 389–412.
(3)   Vesterberg, O., *In "Methods in Microbiology"*, Eds.: Norris, J. R. and D. W. Ribbons, Vol. 5B, Acad. Press, London, 1971, pp. 595–614.
(4)   Vesterberg, O., In "7th International Congress on Chromatography and Electrophoresis", Press. Acad. Europeenes, Brussels, 1973.
(5)   Rilbe, H., Ann. New York Acad. Sci. *209*, 11 and 80 (1973).
(6)   Catsimpoolas, N. and J. Wang, Anal. Biochem. *44*, 436 (1971).
(7)   Catsimpoolas, N. and B. E. Campbell, Anal. Biochem. *46*, 674 (1972).
(8)   Vesterberg, O., Biochim. Biophys. Acta. *257*, 11 (1972).
(9)   Smith, I., P. J. Lightston and J. D. Perry, Clin. Chim. Acta *35*, 59 (1972).
(10) Vesterberg, O. and R. Eriksson, Biochim. Biophys. Acta. *285*, 393 (1972).
(11) Catsimpoolas, N., Separ Sci. *5*, 523 (1970).
(12) Skude, G. and J.-O. Jeppsson, Scand. J. clin. Lab. Invest. 29, Suppl. *124*, 55 (1972).

(13)   Dale, G. and A. L. Latner, Clin. Chim. Acta. *24*, 61 (1969).

(14)   Kenrick, K. G. and J. Margolis, Anal. Biochem. *33*, 204 (1970).

(15)   Söderholm, J., P. Allestam and T. Wadström, Febs lett. *24*, 89 (1972).

(16)   Vesterberg, O., Biochim. Biophys. Acta. *243*, 345 (1971).

(17)   Kjellin, K. G. and O. Vesterberg, 20th Nord. Neurol. Congr., Oslo 1972, To be published in more detail elsewhere.

(18)   Delmotte, P., Z. Klin. Chem. u. klin. Biochem. *9*, 334 (1971).

(19)   Pedersen, L. R., Clin. Chim. Acta., *29*, 101 (1970).

(20)   Dreusler, E., L. Hemmingsen and L. R. Pedersen, Danish Medical Bulletin *19*, 99 (1972).

(21)   Drysdale, J. W., P. Righetti and H. F. Bunn, Biochim. Biophys. Acta. *229*, 42 (1971).

(22)   Finlaysson, G. R. and A. Chrambach, Anal. Biochem. *40*, 292 (1971).

(23)   Righetti, P. and J. W. Drysdale, Biochim. Biophys. Acta. *236*, 17 (1971).

(24)   King, E. E., J. Chromatogr. *53*, 559 (1970).

(25)   Dirksen, M. L. and A. Chrambach, Separ. Sci. *7*, 747 (1972).

(26)   Jeanson, S., T. Wadström and O. Vesterberg, Life Sciences Vol. 11, Part II, 929 (1972).

(27)   Rittner, Ch. and B. Rittner, Z. klin. Chem. u. kon. Biochem. *9*, 503 (1971).

(28)   Vesterberg, O., In "Protides of the Biological Fluids", Ed. H. Peters, Vol. 17, Pergamon Press Oxford, 1970, pp. 383–387.

(29)   Vesterberg, O., Acta. Chem. Scand. *21*, 206 (1967).

(30)   Wrigley, C. W., Biochem. Gen. *4*, 509 (1970).

(31)   Macko, V. and H. Stegeman, Hoppe-Seyler's Z. Physiol. Chem. *350*, 917 (1969).

(32)   Wadström, T. and C. Smyth, to be published in *"Isoelectric Focusing"*, Ed. Arbuthnott, J. P. Butterworths, 1974.

(33)   Vesterberg, O., Separation and characterization of proteins by isoelectric focusing in polyacrylamide gels – A potential tool in clinical chemistry. Abstract 8th Intern. Congr. Clin. Chem., Copenhagen, 1972. Scand. J. Clin. Lab. Invest. *29*, Suppl. 126, 26.1 (1972).

4.2 Isoelectric Focusing Using Inorganic Acidic Ampholytes*

Peter Pogacar and *Richard Jarecki*

Despite excellent separations and widespread applicability, different authors (1, 2, 3, 4) have noted extra bands, inconstant band positions, and other disparities during isoelectric focusing (IF). While using IF for hemoglobin separations and for enzyme differentiation of skin wound extracts (5) we observed a number of bands whose significance is as yet unexplained. Also, during fractionation of, for example, β-naphtyl-acetate disassociating

* Dedicated to Professor Dr. Hans Klein in honor of his 60th birthday.

esterases, up to twenty components appear. Isolation of single protein bands and refocusing these shows further fractionation into as many as seven of the original components.

Different explanations have been offered for these phenomena such as, for example, the following: the presence of artifacts due to monomer acrylic acid in the polyacrylamide gel (PAG) which reacts with protein; oxidation products or heat polymerization taking place during the electrophoretic run; or the origin of chelate-like complexes between the ampholytes and proteins.

The first two reasons could be eliminated with careful technique but if in fact chelate-like complexes between the ampholytes and proteins do take place, then it might be possible to take advantage of this phenomenon and use it as an additional means of protein differentiation. We could, for example, obtain an idea about the distances between the acid and basic groups of the protein because we could react it with a specific ampholyte of known chemical constitution (known distance between acid and basic group of the ampholyte molecule). Only if these molecular distances (determined by the position at which C atoms the acid and basic groups are located) are similar, will reaction take place.

During IF it was possible to change the band positions by changing the voltage or the duration of the run even after the band focusing was complete. This was observed even without the presence of foreign ions. We felt that the reasons for this might be similar to those of the refocusing problem mentioned above and that the key to the solution might be the same.

We decided to synthesize ampholyte mixtures of chemical constitution differing from those now in use. These new ampholytes should enter into different complex bonds with the protein or else form no bonds at all. Comparative refocusing of the bands using thes newly synthesized ampholytes on the one hand, and the standard ampholytes on the other, might clarify the processes taking place.

Svenson's pioneering work of the 1960's on natural pH gradients made possible the widespread interest in isoelectric focusing which followed. He used peptides obtained from the hydrolysis of hemoglobin. (6) Polypeptides, however, had disadvantages such as the conductivity of the different parts of the pH gradient varied too much with consequent variations in heat development; peptides resembled the proteins which had to be separated; and the possibility of peptide breakdown by peptidases.

Vesterberg then succeeded in synthesizing a series of water soluble aliphatic amino-carboxyl acids (7, 8, 9) exhibiting many values of pK for the same type of molecule. For this synthesis (8, 10) a mixture of many different polyethylene polyamines in aqueous solution was reacted with acrylic acid. This introduced a large number of carboxyl groups into the molecules. The resultant mixture contained a great variety of different compounds with molecular weights between 300 and 1000, having different pI values of about 3–10. Moreover, the pH region of 4 to 7, which had been so difficult to obtain, was covered by a large number of these compounds.

For our synthesis we selected as basic carrier tetraethylene pentamine, pentaethylene hexamine, and a low molecular weight polyethylenimine. For the introduction of the acidic groups we used propansultone and Na-vinylsulfonate for the sulfonic, and Na-chlormethylphosphonate for the phosphonic acid.

During the preparative work it was found that propansultone in a basic medium gave the best results for the synthesis of the polyamino-polysulfonic acid mixture. It reacts strongly with primary, secondary, and tertiary amino groups (tertiery amino groups react with propansultone to give betaine-similar compounds).

After some difficulties related to the elimination of foreign ions from these mixtures, solutions of different acidity depending on the acid-base relation were obtained. In order to get single fractions with more defined isoelectric points out of these mixtures, we made use of an electrophoretic separation chamber similar to Vesterberg's (9) with twenty porous separating partitions. We used no water cooling but separated at only 100 V. The temperature did not exceed 40^0 C under these conditions. During the run osmotic volume changes could be seen. After 48 hours no further volume or pH changes were observed. (figure 1)

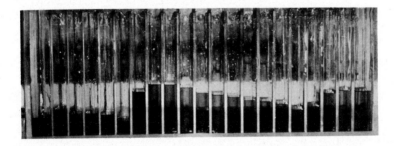

Fig. 1

Surprisingly, it was seen that with an increase in the field strength the individual fractions were separated anew and the pI's of the ampholytes produced were pH 2–3.5 and pH 5.8–9.5. Even with varied synthesis conditions the gap between pH 3.5 and 5.8 could not be bridged. The relatively weakly disassociated 6 basic amino groups of pentaethylene hexamine became neutralized after the introduction of approximately 3 strongly disassociated sulphonic acids; the introduction of a further acid group into the molecule resulted in an acidity jump similar to that seen when a weak base is titrated with a strong acid. Previous quarternization of the amino groups with dimethylsulfate in order to increase their disassociation and their basidity only reduced the gap to 3.8–5.5. Only through the introduction of weakly acid carboxyl groups with acrylic acid in a polyamino-polysulphonic acid mixture with an isoelectric point between 5–8, was it possible to attain an ampholyte mixture with pI's of 3.5–6.

A mixture was made of the individual fractions in order to have a uniform pH gradient which was incorporated as carrier ampholyte in the PAGIF system. A comparison of IF runs with LKB ampholytes on the one hand and the newly synthesized sulphonic acid ampholytes on the other showed a correlation of the main bands. The secondary bands had different patterns dependent upon the ampholytes used. The pattern differences manifested themselves not only in a shift of the bands because of the different distribution of the pI's of the ampholytes, but also different bands manifested themselves in both types.

A determination of the pH gradient in the PAGIF after the run showed no correlation with the pH gradient of the ampholyte mixture before the IF run. (The determination was made by horizontal sectioning of the gel and water extraction of the uniformly-sized segments). It was shown, however, that proper correlation of the gradients could be obtained if the ampholyte separation before the run was done with the same field strength as the IF. This led us to suspect that the ampholytes not only enter into complexes with certain proteins but also that they undergo complexes with one another, apparently at a new pI. The existence of these complexes is dependent upon the field strength during the electrophoresis. This supposition, however, is not consistent with the present electrophoretic theories.

In order to obtain evidence for our assumption we determined the freezing point depression of an electrophoretic mixture. If our supposition were correct we could expect to find less freezing point depression than should be expected according to the ampholyte concentration since freezing point depression is directly dependent upon the amount of particles in solution. For these determinations we used a "half micro osmometer of Knauer" which demonstrates the freezing point depression directly in mille osmols (Table I).

Those separated fractions, as seen in figure 1, were used to determine the dry weight, the pH, and the osmolality (Table 1). As we have mentioned, the originally equal volumes shift according to their concentration so that after equilibrium all cells should contain isotonic solutions. Even though the dry weight shows only relatively small differences, the osmolality of the individual fractions is up to 5 times too low. This would seem to confirm the mentioned multiple linkages. As a marginal note it should be mentioned that a 3.8%

Table 1

Fr.	Vol. ccm	pH	Tr. gew. %	Gefrp. Ern. mosmol
1	35	9,08	11,4	85
2	20	8,44	13,7	40
3	11	8,20	13,0	34
4	6	8,07	12,4	38
5	30	7,77	12,0	42
6	33	7,49	12,0	27
7	10	7,46	11,7	19
8	7	7,39	10,4	18
9	12	7,32	9,9	17
10	14	7,29	9,9	18
11	20	7,26	9,4	20
12	18,5	7,24	9,8	21
13	21,5	7,22	9,9	19
14	18,5	7,13	9,9	34
15	16,5	7,10	10,0	41
16	18	7,09	9,9	43
17	26,5	6,95	9,8	45

(1/6 molar) pentaethylene hexamine solution shows the expected osmolality of 192 m.Osmols.

Unhappily, we must say that we now have more unanswered questions than we had before beginning this investigation. We observed not only the reaction between carrier ampholytes and proteins but also the interaction of the ampholytes wiht each other as well. We believe, however, that the theories concerning the action of carrier ampholytes should be examined more fully. IF, still so problematic at this time, should not only show its great analytical value but also its value in giving information in those cases where interactions between polyvalent carriers of charges play a role. The general signifcance of these speculations might be appreciated if one considers that the biological synthesis of nucleinic acids and proteins and their degradation, as well as the whole complex of enzyme reactions, are a function of the interaction of polyvalent charge carriers.

References

(1)   Percival et al., Aust. J. Exp. Biol. Med. Sci. *48* (2), 171–178 (1970).
(2)   Geithe et al., Biochem. Biophys. Acta *208,* 157–159 (1970).
(3)   Frater, Analytical Biochem. *38,* 536–538 (1970).

(4)   Felgenhauer et al., Prot. Biol. Fluids *19*, 575–578 (1971).
(5)   Jarecki et al., J. Legal Med. *67*, 313–318 (1970).
(6)   Svensson, H., Arch. Biochem. Biophys., suppl. *1*, 132–138 (1962b).
(7)   Vesterberg, O. and H. Svensson, Acta Chem. Scand. *20*, 820–834 (1966).
(8)   Vesterberg, O., British patent No. 11068 18, July 17, 1968.
(9)   Vesterberg, O., Karolinska Institutet, Stockholm, Inaugural Dissertation (1968b).
(10)   Vesterberg, O., Acta Chem. Scand. *23*, 2653–2666 (1969).

4.3 Analytical Gel Isotachophoresis with Ampholine Spacers

Ann L. Griffith and *Nicholas Catsimpoolas*

Isotachophoresis is an electrophoretic separation method based on the Kohl-rausch regulating function (1). It involves the separation of ion species which have the same sign of charge (positive or negative) and all have a common counterion. The ions to be separated are spaced into consecutive zones between the leading and terminating ions according to their constituent mobilities. Because of a fixed concentration ratio, on each side of the moving boundary, sharp boundaries are formed between consecutive zones at equilibrium.

Ornstein (2) and Jovin (3) have discussed the theory of the steady state stacking for the fractionation of proteins. Although, concentrated protein zones can be obtained during stacking it is difficult, in practice, to detect the separation achieved. At present, no detectors are available which can distinguish between consecutive protein zones in a stack. The problem of detection, however, can be solved by spacing the protein zones with carrier ampholytes. This approach to isotachophoresis with spacers has been explored by Svendsen and Rose (4), Routs (5) and also by the work of Griffith and Catsimpoolas (6–8). These investigators have shown the usefulness of Ampholine in isotachophoretic protein separations and also have indicated some of the general factors that affect this type of fractionation. In the present paper an attempt will be made to describe the general experimental conditions for analytical isotachophoresis in polyacrylamide gels using Ampholine spacers.

Experimental

Gel Formulations

A stock acrylamide solution is prepared which contains 29 g acrylamide; 1 g BIS and is made up 100 ml with water. The stock catalyst solution is pre-pared by dissolving 4 mg riboflavin in 100 ml water. The stock pH 6.1 gel buffer consists of 5.2 g TRIS; 39 ml 1M H_3PO_4; 0.5 ml TEMED and is made up to 200 ml with water. A stock sucrose solution is 25% in water. The terminal buffer consists of either 35.8 g beta alanine or 30 g glycine, adjusted to pH 8.2 with TRIS base and made up to a total volume of 2 liters with water osher buffer formulations have been described elsewhere (4). The protein solutions of proper concentration are dissolved in the terminal buffer. The leading electrolyte consists of the stock gel buffer (without TEMED) diluted to the same concentration as the gel buffer actually used in the experiment. The gel solutions are prepared by mixing 3 ml, each, of stock acrylamide solution, the gel buffer, and the stock sucrose solution. Twelve ml of water are added to 12 ml of the above mixture to produce gels of 3.75% acrylamide concentration.

Isotachophoretic Analysis

Glass tubes (15 cm in length and 0.5 cm I. D.) are filled to a height of 13 cm with the gel solution. The mixtures are layered with the gel buffer to produce a flat surface and are photopolymerized for one hour. After polymerization, the tubes are placed in the Canalco electrophoresis apparatus and a mixture of carrier ampholytes, protein sample and stock sucrose solution are layered between the gel-terminating buffer interface. The upper bath (cathode) is filled with the terminal buffer and the lower bath (anode) with the gel buffer (leading electrolyte). Isotachophoresis is carried out with a constant current power supply, usually at 2 mA per gel column for 2 to 3 hours. After isotachophoresis, the gels are removed from the glass tubes and placed for 15 minutes in a staining solution containing 0.5 g bromophenol blue, 50 g mercuric chloride made up to a volume of 500 ml with 50% ethyl alcohol. Destaining is accomplished overnight in a 30% ethyl alcohol. 5% glacial acetic acid solution. An alternate coomassie blue staining procedure as described by Vesterberg (9) has also been used.

Results and Discussion

The effect of Ampholine in increasing the zone length of a protein subjected to isotachophoresis is shown in figure 1. Ovalbumin, at different concentrations, without the presence of ampholytes forms a compact band. In the presence of Ampholine, the band (protein zone), is spread but not in proportion to the amount of the protein.

The length of a protein zone in isotachophoresis is affected by the protein concentration in the zone (as shown on figure 1) and also by the width of the stack if the protein diffuses intro a spacer Ampholine zone of similar mobility. The width of a stack is affected by the concentration of the leading ion, the

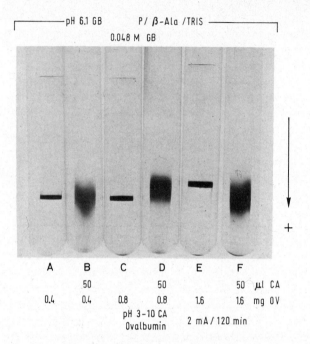

Fig. 1. The effect of Ampholine on the zone length of different concentrations of ovalbumin.

mobilities of the leading and terminating ions, the pH of the leading electrolyte, the protein concentration, and the spacer concentration. Usually, the width of the stack determines the length of the protein zones in a more critical way than the protein concentration in the separation system.

The ability of ampholytes to act as spacers in the separation of protein zones is demonstrated in figure 2. Ovalbumin and beta lactoglobulin were not separated in the absence of carrier ampholytes, in an isotachophoretic system containing phosphate as the leading ion, beta alanine as the terminating ion and TRIS as the counterion. In the presence of ampholyte of different pH ranges, separation of the different forms of beta lactoglobulin and ovalbumin has been achieved. The resolution and the stack width depend heavily on the pH range of ampholytes used. It has been shown (10) for example, that by using the pH 3–10 carrier ampholytes, resolution increases with the amount of ampholyte present in the system. The stack width, however, is very compact, and the protein zones appear thin. Using pH 3–6 or pH 5–8 ampholytes the width of the stack is increased and also, as a consequence, the length of the protein zone has been increased. However, the best resolution was achieved

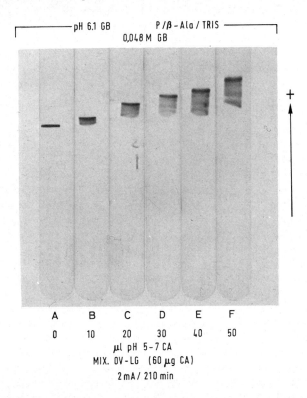

Fig. 2. The effect of different concentrations of pH 5−7 carrier ampholytes on the separation of ovalbumin and beta lactoglobulin.

by using pH 5−7 carrier ampholytes (Figure 2) for this particular separation problem. It should be noted that when pH 7−10 ampholytes were used no separation could be observed. This indicates that the pH 7−10 ampholytes could not be used efficiently in the separation of ovalbumin from lactoglobulin.

An important aspect of these experiments is the demonstration that spacing between the protein zones is roughly proportional to the ampholyte concentration (Figure 2). This can be observed only if the system is at equilibrium; i. e. steady state conditions have been reached after adequate time of electrophoresis was allowed.

Another factor which affects isotachophoretic separations is the pH of the leading electrolyte (Figure 3). The pH of the leading electrolyte can affect separations in two possible ways: (1) by altering the mobility of the leading ion which is pH dependent resulting in faster or slower speed of separation; and (2) by altering the stack width between the leading and terminating ions.

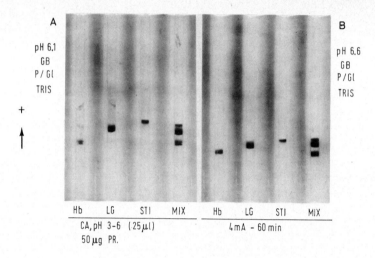

Fig. 3. The effect of changing the pH of the gel buffer (leading electrolyte) on the separation of a mixture of hemoglobin, beta lactoglobulin, and soybean trypsin inhibitor.

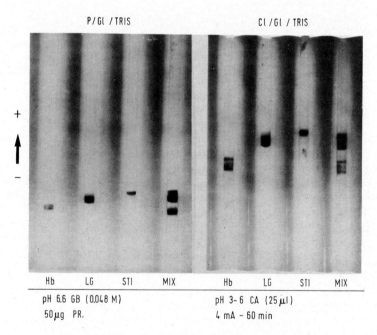

Fig. 4. The effect of the nature of the leading ion (in this case phosphate and chloride) in the separation of hemoglobin, beta lactoglobulin, and soybean trypsin inhibitor.

Two other parameters that can affect the isotachophoretic separation of proteins are the concentration and the nature of the leading ion. An example of using phosphate and chloride as leading ions to separate a mixture of three proteins is shown in Figure 4. Lowering the concentratiun of the leading ion can have a two-fold effect on the isotachophoretic separation. Firstly, wider zones are obtained because the concentration (and therefore the length) of each ampholyte and protein zone are adjusted to the concentration of the leading ion. Secondly, because of the lower conductivtity of the leading electrolyte under constant current conditions, the field strength increases, which results in faster separation. However, too great an increase in field strength will result in heating effects; especially at the zones closer to the terminating ion, since the conductivity decreases from the leading electrolyte to the terminating ion.

Another important factor in isotachophoresis is the separation time. Adequate time should be allowed for the system to reach equilibrium. In contrast to zone electrophoresis, where diffusion and thus zone widening increase with time leading to a loss in resolution, in isotachophoresis there is an absence of diffusion because it is counteracted by the potential gradient between zones. In isotachophoresis improved separations can be achieved by allowing the system to reach the steady state.

An example of the application of isotachophoresis in the separation of several fractions of human serum proteins is shown in figure 5. While no fractionation occurs without the presence of ampholytes, components are readily resolved in the presence of pH 5–8 carrier ampholytes. The overall separation pattern can be considered better than that of simple zone electrophoresis, but inferior to that obtained by disc electrophoresis. This is probably due to the lack of sufficient number of components of intermediate mobility in the ampholyte mixture.

The appropriate strategy for the separation of an unknown mixture of proteins by isotachophoresis will involve the following: Choice of the polyacrylamide gel concentration so that sieving effects can be avoided; choice of the appropriate leading ion, terminating ion and counterion; choice of the concentration of the leading ion in the pH of the leading electrolyte; demonstration of stacking of the protein of interest in the absence of carrier ampholytes under the chosen isotachophoretic separation conditions; choice of the pH range of the carrier ampholytes, and determination of the amount of ampholytes present in the sample for optimum spacing and separation. The synthesis of new spacers will certainly expand the capabilities of this technique. At present, the Ampholine carrier ampholytes, although not ideally suited for isotachophoresis, offer an experimental tool for studying the spacing effect in protein separation.

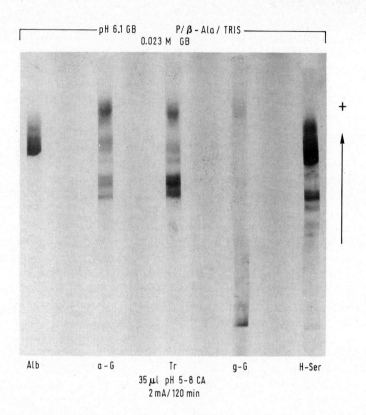

Fig. 5. Isotachophoretic separation of human serum protein fractions in the presence of pH 5−8 carrier ampholytes.

References

(1)   Kohlrausch, F., Ann. Phys. Leipzig *62,* 209 (1897).

(2)   Ornstein, L., Ann. N.Y. Acad. Sci. *121.* 321 (1964).

(3)   Jovin, T., Ann. N.Y. Acad. Sci. *209,* 477 (1973).

(4)   Svendsen, P. J. and C. Rose, Sci. Tools *17,* 13 (1970).

(5)   Routs, R. J., Electrolyte Systems in Isotachophoresis and Their Application to some Protein Separations. Dissertation, Technological University, Eindhoven, (1971).

(6)   Griffith, A. and N. Catsimpoolas, Anal. Biochem. *45,* 192 (1972).

(7)   Griffith, A., N. Catsimpoolas and J. Kenney, Ann. N.Y. Acad. Sci. *209,* 457 (1973).

(8)   Catsimpoolas, N. and J. Kenney, Biochim. Biophys. Acta *285,* 287 (1972)

(9)   Vesterberg, O., Biochim. Biophys. Acta *257,* 11 (1972).

(10) Griffith, A. L. and N. Catsimpoolas (in preparation).

Abbreviations

Hb, Hemoglobin; LG, beta lactoglobulin; STI, soybean trypsin inhibitor; CA, carrier ampholytes; Pr, protein; GB, gel buffer; P, phosphate; Gl, glycine; Cl, chloride; TEMED, N.N.N′, N′, -tetramethylenediamine; BIS, N, N′-methylenebisacrylamide; TRIS, 2-amino-2-(hydroxymethyl)-1,3-propanediol; MIX, mixture; B-Ala, beta alanine; OV, ovalbumin; TB, terminal buffer; Alb, albumin; A-G, alpha globulin; Tr, transferrin; g-G, gamma globulin; H-Ser, human serum.

Chapter 5. Methods of Quantitative Pattern Evaluation

5.1 Some Prerequisites for Quantitative Micro-densitometry[1]

Robert C. Allen

Methods for the quantification of polyacrylamide gel pherograms and zymo-grams depend on the use of various types of high resolution microdensito-meters coupled to some type of integrator. The desirable characteristics and requirements for such instruments have been previously set forth at length (1–5).

Whereas in the past such instrumentation was marginally acceptable in its capability to measure and resolve 30 to 50 micron-wide bands, present available instrumentation appears to adequately meet this requirement. With the exception of densitometers designed to quantify nucleic acids in the 260 nm region or those measuring fluorescence, either induced such as tryptophane phenylalanine and tyrosine excitation, or bound fluorescent dyes such as fluorescamine, quantification relies mainly on measuring the adsorp-tion of visible light by protein stains; or, in the case of isozymes, an insoluble complex of stain and released product. Thus, a major prerequisite for accurate reproducible quantification lies in the precision of the techniques employed prior to and during the separation.

For both protein and isozyme techniques not only should a determined amount of material be utilized in the separation, but gel %T, C, temperature and run time must be accurately controlled to reproduce qualitative and quantitative data. Also, staining procedures should be precisely reproduced. In the cases where destaining procedures are utilized, as in the example of proteins, electrophoretic destaining may not give similar results to destaining by diffusion. Since many stains and stain-product complexes may fade with storage or are light sensitive, care must be taken to perform densitometry as soon as possible after the background dye is removed both to maintain similar conditions of stain intensity and gel swelling or shrinkage resulting from stain procedures. This is particularly important when Amido Black is used as a protein stain (4). With isozymes, densitometry should be performed as soon as possible after the reaction is stopped, and in all cases a standard sample

---

[1] Dedicated to Professor Hans Klein in honor of his 60th birthday.

should be included to control dye lot and reagent variations. In many instances where resolution is adequate, colored film may also be utilized to take pictures under standard lighting conditions of the separation for both permanent records and densitometry. The last frame on a film may be used as a reference standard for the first frame on the succeeding film to adjust for film and processing differences.

The increasing use of isoelectric focusing with its very high resolution potential has further complicated the quantification of electropherograms. Where, for example, one found 20–25 bands in PAGE, the same sample may show up to 50 protein bands with isoelectric focusing or additional finely resolved multiple molecular forms of an enzyme. Similarly, SDS systems may show 50 or more bands.

An approach to improve quantitative densitometric evaluation of these systems, in addition to carefully controlled procedures, is to reduce the lensing and flare effects of the gel itself. Obviously from a geometrical standpoint, using a flat gel offers advantages over cylindrical gels for this purpose (4).

Secondly, a thin gel is more desirable than a thicker gel, both from the reduced lens effect and reduced background. Thus, it would seem desirable to utilize the thinnest slab gel consistent with ease of manipulation and accurate sample application. Gel slabs between 0.6 mm and 1 mm thick appear then to offer a practicable compromise for this purpose since standard microsyringes may be utilized.

In figure 1, a comparison of the densitometric trace of a 1 mm and 3 mm gel are shown. It is evident that the resolution in the GC region is better in the thin gel and that less background is present in the diffuse gamma region (4). A more striking example of gel thickness is shown in figures 2A and B which compare skin proteins and esterases separated by isoelectric focusing. In the thicker gel (figure 2) pH gradient diffusion and concomitant band diffusion following electrophoresis during enzyme reaction time tend to obscure the bands, while on the corresponding acid fixed and stained portion of the gel the protein bands are sharply resolved. On the other hand, the thin gel (figure 2B) in which the enzyme reaction is more complete shows much sharper resolution.

Further, the thinner the gel in the case of isozyme studies, the more complete the reaction that occurs in a specific time interval due to the decreased diffusion path of substrate and complexing agent to the enzyme. Thus, for example, in a 3 mm thick gel esterases are stained to a depth of 0.4 mm during the first 15 minutes of reaction which is still in the linear range of a Lineweaver-Burk plot. Thus, in a 3 mm thick gel less than 30%, the enzyme has reacted with substrate and complexing agent while in a 0.7 mm gel in the same reaction time, practically all of the enzyme should be reacted.

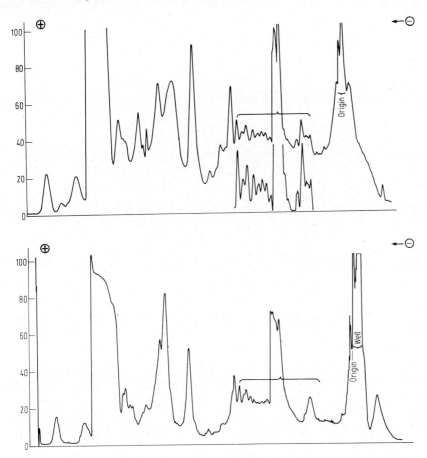

Fig. 1. Human serum protein separated in a tris-citrate-borate discontinuous buffer system at pH 9.0. A step gradient gel of 3.0, 6, 8, 12 per cent T and 3.5 per cent C was used. Gels were stained with Coomassie Brilliant Blue R 250 and scanned on an ORTEC model 4300 integrating microdensitometer. A. 3 mm gel. B. 1 mm gel.

Similarly, thin slab gels for SDS separations are more desirable for densitometry and subsequent drying and autoradiography when correlation of radioisotope binding with specific protein bands is desired.

Thus, for accurate reproducible quantification of proteins, enzymes, nucleic acids and denatured proteins separated by PAGE or PAGIF a major consideration is in the precision of the sample preparation, separation, and staining techniques as well as in the choice of an appropriate microdensitometer. The requirements listed here for quantitative purposes apply as well to the improvement of results wherein only qualitative findings are sought.

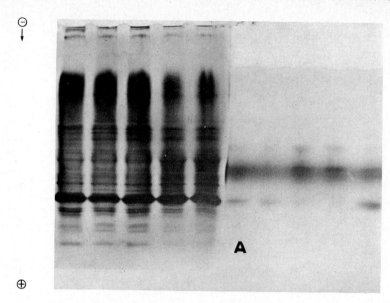

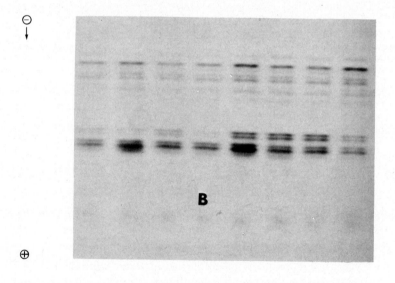

Fig. 2. Isoelectric focusing of human skin proteins and esterases extracted with phenoxyethanol. 2A protein and esterase separation on 3 mm thick gel slab, $\frac{1}{2}$ gel stained for protein with Coomassie Brilliant Blue R 250 and other half for nonspecific esterase activity with alpha-naphthyl acetate as substrate and Fast Red TR complexing agent. 2B same separation on 1 mm gel of esterases similarly reacted as in 2A.

References

(1)   Hansl, R., Ann. N. Y. Acad. Sci. *121*, 391 (1964).
(2)   Allen, R. C. and G. R. Jamieson, Anal. Biochem. *16*, 450 (1966).
(3)   Hannig, K., Z. Analyt. Chem. *243*, 522 (1968).
(4)   Maurer, H. R. and R. C. Allen, Clin. Chim. Acta. *40*, 359 (1972).
(5)   Kling, O., This volume, chapter 5.2.

5.2 Problems of Quantitative Evaluation of Disc Pherograms

O. Kling

By means of gel electrophoresis significant biochemical compounds can be separated into their components with remarkably high resolution. However, this qualitative fractionation also calls for the quantitative determination of the identified components. In this instance it is less important whether the relative or the absolute portions of the fractions are determined. Photometric evaluation (densitometry) has been selected as method for several reasons which need not be explained here. The following is a concise description of the special characteristics of gel electrophoresis which determine the instrumentation and are realized in the disc electrophoresis attachment ZK 4 for the PMQ spectrophotometer[1].

Application of the Lambert-Beer-Bouguer law

$$E = \log \frac{I_0}{I} = \log \frac{1}{\tau} = \epsilon_\lambda \cdot c \cdot d \tag{1}$$

is a precondition for quantitative measurement. It establishes the quantitative relationship between the parameters τ or E (transmittance or absorbance) and concentration c. Transmittance τ is the primary parameter which is converted into absorbance values by suitable means (internal transmittance-absorbance conversion, logarithmic recording).

Layer thickness d

The high resolving power of the pherograms yields clearly separated fractions with widths and distances in migration direction down to 50 microns. Optical resolution better than 50 microns is therefore required. The layer thicknesses

---

[1]   Manufacturer: CARL ZEISS, D-7082 Oberkochen, West Germany.

of flat and cylindrical gels lie generally between 1 and 7 mm, and small apertures are required in order to achieve the necessary scanning resolution.

Stops in the ray path alone cause high energy losses and unfavorable signal-to-noise ratios. Better conditions are achieved with an achromatic mirror system to produce in the sample an image of the continuously variable measuring slit. The layer thickness of cylindrical gels penetrated by the light is not defined. Special cells deform cylindrical gels of 3, 5 or 7 mm diameter under light pressure without mechanical destruction into plane-parallel layers of 2.3, 4.2, or 6.2 mm diameter. For quantitative determinations flat gels can be fixed in the ray path in a special guide with precision drive. The theoretical resolving power of 20 microns can be fully utilized only when the preparation fulfills the following requirements:

1.1 the fraction bands are not curved;
1.2 the fraction bands are at right angles to the migration direction.

Extinction coefficient ϵ_λ

Measurement must be possible at any wavelength including UV (native absorption of nucleic acids and aromatic amino acids at 260 and 280 nm respectively). Direct measurement in the absorption maximum of the sample is particularly specific and leaves the protein fractions unchanged. Staining and simultaneous denaturation and fixation result in higher absorption coefficients of the adducts and in higher measuring sensitivity, with the following restrictions:

2.1 Compared with the stains the protein fractions have different staining coefficients;

2.2 The staining is not homogeneous and quantitative (e. g. staining only of the outer zone of cylindrical gels);

2.3 Destaining of substance-free zones is incomplete and inhomogeneous.

Absorbance E

The measured absorbance $E(x)$ as a function of the scanning coordinate x consists of several portions

$$E(x) = E_S + E_G' + E_B \tag{2}$$

E_S is the absorbance of the substance to be determined, from which conclusions are to be drawn to the concentration by means of (1). E_G' and E_B are the unspecific basic absorbances of gel and buffer which interfere especially in the UV. High basic absorbances reduce the attainable measuring accuracy and impair the signal-to-noise ratio and thus resolution and detection limit.

The apparent total absorbance E_G' of the gel consists of two portions

$$E_G' = E_G + E' \tag{3}$$

where E_G is the specific absorption of the gel which is determined above all by the purity of the monomers used. E' is a spurious absorbance caused by light scattering in the gel. As the scattering is indirectly proportional to λ^4, quantitative direct measurements in the UV may be impaired. Inhomogeneities in the gel layers may result in markedly location-dependent background effects.

The above parameters are essentially dependent on the preparation and can be overcome by suitable standardization of the working procedures. Elimination of the scattering is especially important since the Lambert-Beer-Bouguer law (1) applies only to optically clear substances.

Integration

Quantitative statements require measurement of the absorbance as a function of the scanning coordinate x followed by integration

$$c \sim \int_{x_1}^{x_2} E(x)\, dx \tag{4}$$

The measured absorbance $E(x)$ as a function of the scanning coordinate x recorder, computing recorder, computer) depends on the safety with which the boundary conditions can be defined which determine the measuring result.

The integration boundaries are determined by the preparation and the photometrically obtainable scanning resolution. In case of overlapping bands additional criteria must be determined by systematic measurements to establish the integration ranges.

The value of the integral is determined by the integration boundaries and the properties of the base line (position, slope, bent). It must be proved by systematic examinations whether the properties of the base line are constant, additive components or variables. Sufficient numerical results are not yet available. To illustrate the problems of quantitative evaluations the essentials of the computer program of the automatic pherogram photometer KFE 4[2] are listed here:

a) Subdivision of the variable measuring track into 200 equidistant measuring points.

b) Determination of the absorbance values of start and end points, linear base line correction.

c) Averaging from several measuring results at each measuring point (number of measuring values selectable as program parameter).

d) Determination of the integrating boundaries by computation of the genuine absorbance minimum between two fractions each (minimum absorbance difference selectable as program parameter).

e) Elimination of minimum fractions, e. g. simulated by artifacts (minimum fraction value selectable as program parameter).

f) Determination and printout of relative fraction points according to the above criteria.

It is possible to produce computer programs for the quantitative evaluation of disc pherograms, provided all criteria for a judgement are known. Such criteria are, however, available only for a few specific problems.

5.3 Transient State Isoelectric Focusing in Poly-acrylamide Gel (Scanning Isoelectric Focusing)

Nicholas Catsimpoolas

Introduction

The need for an analytical micro-method involving isoelectric focusing (1,2) led to the development of the *gel technique* of which several variations have been described in the literature (3). The advantages of gels as anticonvection media combined with shorter focusing times and conservation of sample and expensive carrier ampholytes made this method a valuable new tool for the analysis of proteins. In addition to the above, the incorporation of a direct optical scanning system (6) applied *in situ* (i. e. without interruption of the current) expanded the capabilities of this technique because the separation path can be continuously monitored and the relevant experimental parameters can be acquired and processed on-line in digital form (7). Thus,

---

[2] Manufacturer: CARL ZEISS, D-7082, Oberkochen, West Germany.

information storage, retrieval and reporting of data can be considerably facilitated. Furhtermore, physical constants such as the isoelectric point, the apparent diffusion coefficient, and the slope of the pH-velocity curve at the isoelectric point (pE) can be computed from kinetic data.

Transient state isoelectric focusing (TRANSIF) involves three stages from which kinetic data can be collected. These are: (1) focusing in the presence of an electric field until a steady-state is reached; (2) defocusing in the absence of the electric field; and (3) refocusing by re-applying the electric field. During the refocusing stage the protein ideally should exhibit a Gaussian concentration distribution at all times.

Preliminary Considerations

Isoelectric focusing has been considered to be an *equilibrium* method, that is each species seeks an *equilibrium* position along the separation path which is the result of electrical and pH gradients. Any deviation from the *equilibium* point (e. g. by diffusion) will activate a restoring force (electrical transport) which tends to keep the species focused in a narrow zone. Svensson (8, 9) expressed the concentration distribution of an ampholyte at the point of focusing as a Gaussian curve with inflexion points at:

$$\sigma = \pm \sqrt{\frac{D}{pE}} \tag{1}$$

where:

$$p = -\frac{du}{d(pH)} \cdot \frac{d(pH)}{dx} \tag{2}$$

and σ denotes the standard deviation of the Gaussian distribution; D, the diffusion coefficient; E, the field strength; $d(pH)/dx$, the pH gradient; and $- du/d(pH)$, the slope of the pH-mobility curve at the isoelectric point. Since the parameters D and $du/d(pH)$ are fixed by the nature of the protein, constant values of E and $d(pH)/dx$ should produce a zone which at *equilibrium* will remain in focus indefinitely. However, the concentration distribution of a focused zone as described in Eq. 1 depends on a number of secondary factors (7) as illustrated in figure 1. Most of the relevant experimental parameters (with the exception of viscosity and conductance) corresponding to the steady-state can be measured experimentally with the scanning isoelectric focusing (SCIF) technique in polyacrylamide gels. Thus, conditions involving pH gradient instability, zone displacement, zone area instability and other transient phenomena (7, 10) can be evaluated in a quantitative manner. In addition, other important parameters such as the

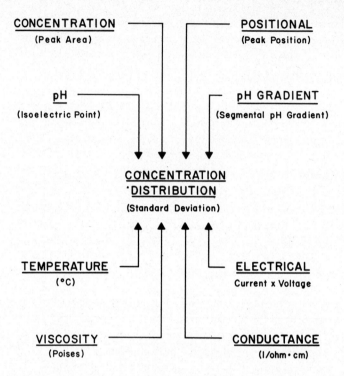

Fig. 1. Schematic diagram of secondary factors affecting the concentration distribution of a focused zone and their measurable relevant parameters.

resolving power, resolution, isoelectric point, apparent diffusion coefficient, and the apparent slope of the pH-velocity curve of a protein at its isoelectric point can be estimated.

The Scanning Assembly and Procedure

The basic scanning unit includes the Gilford 2000 recording spectrophotometer, a Beckman DU monochromator, a Gilford Model 2410-R Synchronous Linear Transport System, a Gilford Model 410 Digital Absorbance Meter, an Infotronics Model CRS-208 Automatic Digital Integrator equipped with data printer, a Gelman constant voltage power supply, a milliammeter, and a Cramer elapsed time indicator. The aperture plate (0.1 mm) is supplied with the Linear Transport System. The devices for holding the quartz tubes (Amersil, Inc.) either in the horizontal or vertical position have been described in detail elsewhere (6, 11).

The polyacrylamide gels prepared by photopolymerization contain 0.4 g acrylamide, 15 mg N,N',-methylenebisacrylamide, 0.5 ml (2% final concentration) of 40% stock Ampholine (LKB) solution, 0.65 ml of 0.004% solution of riboflavin in water, and are made up to 10 ml with water. The test protein is included in the gel formulation. The anolyte and catholyte are 0.01 M phosphoric acid and 0.02 M sodium hydroxide, respectively. Focusing is carried out initially at 50 Volts for 30 minutes then at 200 Volts for the duration of the experiment. Some experiments (whenever indicated) were carried out with 6, 8, 10, and 12%T and 5.4%C formulations[1].

Scanning is carried out at regular time intervals and the following experimental data are recorded; the total elapsed time after the current is turned on; the elapsed scanning time of each peak given by the integrator; the maximum absorbance of each peak; the integrated peak area given by the integrator; and the current. From these data relevant parameters can be calculated as shown below.

Derivations of Relevant Parameters

The standard deviation of the concentration distribution of a focused zone is obtained by:

$$\sigma = \frac{\text{Peak Area}}{\sqrt{2\pi}\,\text{Peak Height}} \tag{3}$$

Experimentally, σ is measured by:

$$\sigma = \frac{C\,v\,F}{2.5 \times A\,f\,R} \tag{4}$$

where σ is the standard deviation of the Gaussian distribution (in cm), A is the absorbance at peak maximum (in optical density units), F is the full scale absorbance (in optical density units), f is the count rate of the digital integrator (in counts/mV · sec), C is the total area count (in counts), v is the scanning rate of the Linear Transport (in cm/sec), and R is the recorder millivoltage (or photometer output) corresponding to 100% full scale (in mV).

The peak postion in the separation path is measured by:

$$\bar{x} = v(t_p - t_r) \tag{5}$$

---

[1]   C = % crosslinking (ratio of BIS (w/v) to total T)
      T = total gel concentration (acrylamide + BIS)

where $\bar{x}$ is the peak position (in cm) with zero set at reference mark, t_p is the time of scanning peak maximum (sec), and t_r is the time of scanning reference mark.

The segmental pH gradient is measured using two marker proteins of closely spaced isoelectric points using Eq. 6

$$\frac{\Delta(pH)}{\Delta x} = \frac{pI_A - pI_B}{\bar{x}_A - \bar{x}_B} \qquad (6)$$

where $\Delta(pH)/\Delta x$ is the segmental gradient (in cm^{-1}), pI is the isoelectric point, and (A) and (B) are pI marker proteins. The assumption is made that $\Delta(pH)/\Delta x$ is constant between pI_A and pI_B, where $pI_A - pI_B$ represents a small segment of the total pH gradient in the separation path. Under these conditions the error due to a non-linear pH gradient will be small especially in areas away from the two electrodes where abrupt steepness of the curve may occur.

If a third marker protein is included in the segmental pH gradient, its isoelectric point can be calculated from equation 7.

$$pI_U = pI_A + (\frac{\Delta(pH)}{\Delta x}) (\bar{x}_U - \bar{x}_A) \qquad (7)$$

If the three markers are not at pH equilibrium, the derived isoelectric point value will vary significantly with time.

The peak area (mV · sec) is measured with a digital integrator.

The resolving power at the point of focusing of a zone can be derived from the segmental pH gradient at the standard deviation by:

$$\Delta pI = 3[\Delta(pH)/\Delta x]\sigma \qquad (8)$$

where ΔpI is the minimum pI difference for complete resolution of two focused zones (in pH units).

The resolution Rs of two peaks separated by a distance $\Delta\bar{x}$ is given gy:

$$R_s = \frac{\Delta\bar{x}}{1.5(\sigma_A + \sigma_B)} \qquad (9)$$

where σ_A and σ_B represent the average standard deviation of zones (A) and (B).

The apparent diffusion coefficient (D) is obtained during the defocusing stage by scanning a focused zone in the absence of current at regular intervals from equation 10.

$$\sigma_D^2 = \sigma_F^2 + 2Dt_D \qquad (10)$$

where σ_D^2 is the variance during diffusion, σ_F^2 is the variance at zero diffusion time, and t_D is the elapsed diffusion time. A plot of σ_D^2 (in cm^2) vs. $2t_D$ (in sec) produces experimental points which, when treated by least squares linear regression analysis give D as the slope of the straight line.

The apparent slope of the pH-velocity curve of a protein at its isoelectric point (pE) can be obtained by first focusing the zone maximally, then allowing it to diffuse for some time, and subsequently refocusing it at constant voltage.

The refocusing process is described by Equation 11

$$\ln\left(\frac{\sigma_R^2 - \sigma_F^2}{\sigma_D^2}\right) = -2pEt_R \tag{11}$$

where σ_R^2, σ_F^2, and σ_D^2 is the variance during refocusing, at the end of focusing, and end of diffusion respectively. E is the field strength, t_R is the elapsed time of refocusing, and $p = [du/d(pH)][d(pH)(dx)]$. If the pH gradient, $d(pH)/dx$, and the field strength E at the vicinity of the zone are known, the slope of the pH-mobility curve at pI, $du/d(pH)$, can be estimated from the slope of the linear portion of the plot of Equation 11 by least squares linear regression analysis. The major error in utilizing Equation 11 is the experimental difficulty and uncertainty in measuring the conductance of the zone.

Applications

Scanning Isoelectric Spectra

An example of typical scanning isoelectric spectra is presented in figure 2. The resolution obtained by this method is excellent. The protein zones can become very sharp by increasing the field strength. Thus, the peak height depends not only on the protein concentration, but also on the field strength and the segmental pH gradient at the point of focusing. It is, of course, understood that the fixed parameters such as extinction coefficient at the isoelectric point, diffusion coefficient, and electrophoretic mobility at pI also contribute to the peak height.

The most striking advantage of the scanning technique is that the separation pattern can be visualized continuously during the experiment. The sensitivity of detection at 280 nm suffers because of background noise which cannot be entirely eliminated with single beam instruments as used in these studies. However, the use of a ratio recording system at two wavelengths may considerably improve the lower limits of detection especially at low temperatures and high field strengths.

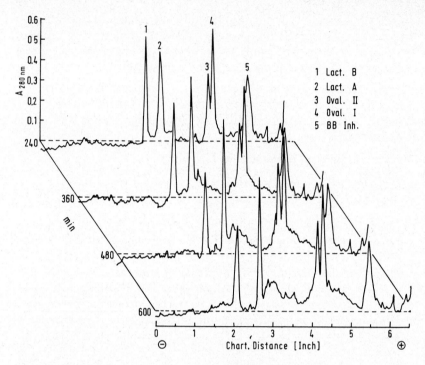

Fig. 2. Scanning isoelectric spectra of beta lactoglobulin, ovalbumin and Bowman-Birk soybean protease inhibitor in the pH 3–7 range. Final ampholyte concentration was 2% and 100 µg of each protein was used for the experiment.

Diffusion in the Absence of the Electric Field

The superior resolution that is obtained with the scanning method, in comparison to staining methods, can be attributed to the presence of electric field during detection. Figure 3 illustrates the magnitude of zone spreading at various time intervals after turning off the current. If the isoelectric focusing experiment is performed at low temperatures, the diffusion spreading can be worse since the staining procedures are usually carried out at room temperature. It is therefore worth mentioning that even densitometry after termination of the experiment will not produce equivalent quality of spectra to those obtained by *in situ* scanning techniques.

Minimal Time of Focusing

An important aspect of isoelectric focusing is the estimation of the minimal time of focusing of a protein zone. This type of evaluation can be facilitated with the scanning technique if the protein is distributed uniformly through-

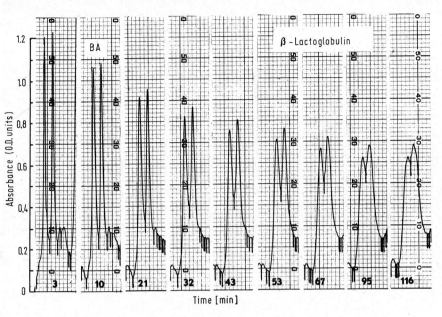

Fig. 3. Zone spreading in the absence of current of isoelectrically focused beta lacto-globulin A and B (100 μg) in the pH 3–10 region.

out the separation path at the beginning of the experiment. The isoelectric point divides the separation path into two regions where the protein will be predominantly positively or negatively charged. These two charged species will migrate continuously toward the isoelectric point and they will accumulate there in a fashion which is basically dependent on the titration curve of the protein. Again, depending on the individual titration curve of the protein, the oppositely charged species may accumulate at the same time maximally at the isoelectric point, or one of the species may accumulate first maximally followed by the other. If one then plots the integrated peak area of the discernable focused zone vs. time, one plateau should be observed in the former case, and two plateaus in the latter. This is illustrated in figure 4 for the focusing of ovalbumin in the pH 3–10 range. The two plateaus can be seen clearly, and in this case, the second plateau indicates the point of minimal time of focusing.

Peak Area Instability and Peak Position Driftage

The plots of the integrated peak area vs. time as described above can be used not only in the estimation of the minimal focusing time, but also in the evaluation of the stability of the area of a zone. It is evident that at the time

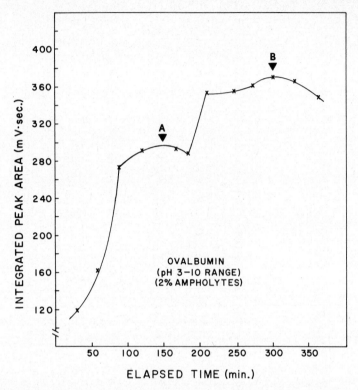

Fig. 4. Plot of integrated peak area vs. time for ovalbumin (100 μg) focused in the pH 3–10 range (2% ampholytes). A and B indicate the positions of the first and second plateaus, respectively.

of maximum focusing the concentration of the protein in the zone should become maximal and remain constant under steady-state conditions. However, it has been observed that the peak area in a focused zone may decline or increase with time. This type of instability in some areas imposes difficulties in evaluating the minimal focusing time and also in obtaining quantitative data.

Another undesirable phenomenon in isoelectric focusing is peak driftage or zone displacement toward one of the electrodes. This positional instability can be evaluated by the scanning technique by plotting the peak position vs. the elapsed electrophoresis time.

Segmental pH Gradient and Isoelectric Point Derivation

The lack of pH gradient *equilibrium* in isoelectric focusing can be evaluated by examination of the change of segmental pH gradients with time. The

assumption is made that after optimum focusing the protein zones are at pH *equilibrium* and that the pH gradient is approximately linear for a small segment of the separation path. An example of such a plot is shown in figure 5. In this particular case the graph indicates that the pH gradient changes progressively to lower values (pH units/cm) with increasing time of electrolysis and that this change is different for two different segments within the same separation path.

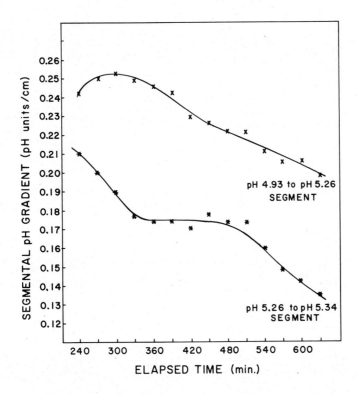

Fig. 5. Demonstration of segmental pH gradient instability using ovalbumin I (pI = 4.93), beta lactoglobulin A (pI = 5.26), and beta lactoglobulin B (pI = 5.34) as pH markers. The protein (100 μg each) were focused in the pH 3−7 range (2% ampholytes).

The isoelectric point of a protein can be calulated by using two pI markers. Table 1 shows such calculations for the isoelectric point of beta lacto-globulin A (pI = 5.26) using beta lactoglobulin B (pI = 5.34) and ovalbumin I (pI = 4.93) as markers. Excellent reproducibility was obtained from two sets of experiments using different concentrations of ampholytes.

Table 1. Derivation of the isoelectric point of beta lactoglobulin A focused in the pH 3–7 range (1:1 mixture of pH 3–6 and pH 5–7 Ampholine).

Total elapsed time (min)	Isoelectric point* (pH)	Total elapsed time (min)	Isoelectric point** (pH)
125	5.256	240	5.255
160	5.235	270	5.250
190	5.288	330	5.243
246	5.249	360	5.241
300	5.236	390	5.245
340	5.252	420	5.246
360	5.234	450	5.250
390	5.248	480	5.251
420	5.249	510	5.252
450	5.249	540	5.247
480	5.249	570	5.243
510	5.239	600	5.242
540	5.239	630	5.239
570	5.237		5.246
	5.243		(± 0.004)
	(± 0.017)		

  * 1% Ampholine, 200 µg protein
** 2% Ampholine, 100 µg protein

Standard Deviation and Resolving Power

Two other important parameters in isoelectric focusing namely, the standard deviation and the resolving power, can be estimated with the scanning technique. Examples of such calculations are shown in table 2. The quantitative data clearly show the high resolving power of the method and the sharpness of the focused zones. Derivation of these parameters is also important for comparative methodological studies of the technique. For example, figure 6 illustrates the effect of electric field strength on the resolving power in the region of focusing of ovalbumin I and beta lacto-globulin B in the pH 4–6 range.

Estimation of Apparent Diffusion Coefficients at pI

The apparent diffusion coefficient of a protein at its isoelectric point can be determined by allowing the focused zone to diffuse in the absence of current. The zone is scanned at regular time intervals and the apparent diffusion coefficient D is obtained as the slope of the plot of σ^2 vs. 2t. Such plots for beta lactoglobulin B, myoglobin and ovalbumin are shown in figure 7. In this particular experiment the proteins were focused in the pH 3–10 range. Their apparent diffusion coefficients at 25°C were estimated by linear regression

Table 2. Derivation of the standard deviation and the resolving power for beta lacto-globulin A focused in the pH 3–7 range (1:1 mixture of pH 3–6 and pH 5–7 Ampholine).

Total elapsed time (min)	Standard deviation* (cm)	Resolving power** (pH units)
240	0.0672	0.042
270	0.0524	0.031
300	0.0491	0.027
330	0.0450	0.023
360	0.0524	0.027
390	0.0586	0.030
420	0.0471	0.024
450	0.0427	0.023
480	0.0523	0.027
510	0.0478	0.025
540	0.0546	0.026
570	0.0542	0.024
600	0.0506	0.022
630	0.0549	0.022
	0.0520	0.026
	(± 0.0060)	(± 0.005)

* 1% Ampholine, 200 μg protein   ** 2% Ampholine, 100 μg protein

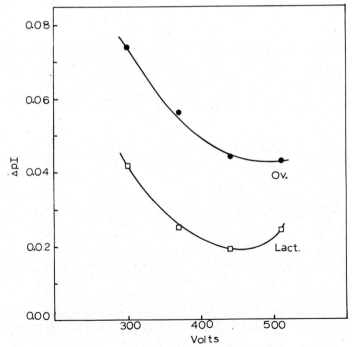

Fig. 6. The effect of voltage on the resolving power in the region of ovalbumin I and beta lactoglobulin B focused in the pH 4–6 range (2% amphylytes).

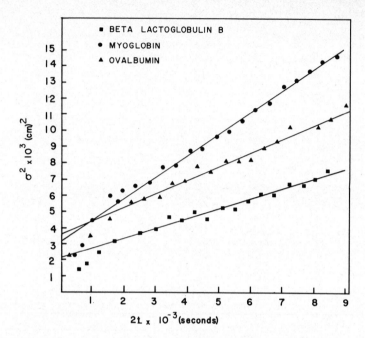

Fig. 7. Plots of σ^2 vs. 2t for estimation of the apparent diffusion coefficient of the indicated proteins at 25°C.

analysis to be as follows: beta lactoglobulin B, 6.19×10^{-7} cm^2 · sec^{-1}; ovalbumin, 8.54×10^{-7} cm^2 · sec^{-1}; and myoglobin, 13.42×10^{-7} cm^2 · sec^{-1}. It should be mentioned that the apparent diffusion coefficients measured by this technique may reflect gel sieving effects, protein concentration effects, possible variation of the viscosity of ampholytes at the region of diffusion, and possible aggregation or denaturation of proteins during migration toward the isoelectric point, or at the isoelectric point.

Kinetics of Refocusing

An example of a plot of $\log\left[(\sigma_R^2 - \sigma_F^2)/\sigma_F^2\right]$ vs. t for the refocusing stage of beta lactoglobulin B focused in the pH 4—6 range is shown in figure 8. The slope of this plot represents the value of 2pE/2.303 which in this case was found to be 2.06×10^{-3} sec^{-1} by least squares linear regression analysis. Assuming a uniform conductivity throughout the separation path, E = 22 V · cm^{-1} and d(pH)/dx = 0.15 cm^{-1}. These data and assumptions yield a value of du/d(pH) = 31.33×10^{-5} cm^2 V^{-1} sec^{-1} for beta lactoglobulin B which is unrealistically high and requires corrections in terms of gel concentration, ampholine concentration and other variables.

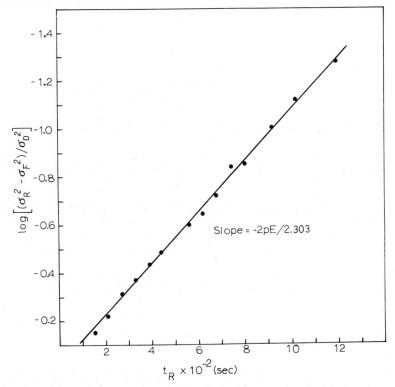

Fig. 8. Kinetic plot obtained from the refocusing of beta lactoglobulin B in the pH 4–6 range.

Concluding Remarks

Some important advantages of the transient isoelectric focusing technique include the following: direct visualization of the separation pattern; continuous monitoring; choice of wavelength; high resolution; direct on-line digital acquisiton and processing of data; and quantitative measurement of relevant parameters and physical constants.

It is conceivable that an automatic multiple-column scanning assembly coupled to a computerized system for data collection and storage and processing may be very useful in routine applications of the technique.

References

(1)   Svensson, H., Acta Chem. Scand. *16*, 456 (1962).
(2)   Vesterberg, O. and H. Svensson, Acta Chem. Scand. *20*, 820 (1966).
(3)   Catsimpoolas, N., Separation Sci. *5*, 523 (1970).

(4) Haglund, H., in "Methods of Biochemical Analysis" (D. Glick, Ed.) Interscience, New York, 1971, vol. 19, p. 1.

(5) Wrigley, C. W., in "New Techniques in Amino Acid, Peptide and Protein Analysis" (A. Niederwieser and G. Pataki, Ed.) Ann Arbor Science Publishers, 1971, p. 291.

(6) Catsimpoolas, N. and J. Wang, Anal. Biochem. *39*, 141 (1971).

(7) Catsimpoolas, N., Anal. Biochem. *54*, 66 (1973).

(8) Svensson, H., Acta Chem. Scand. *15*, 425 (1961).

(9) Svensson, H., Acta Chem. Scand. *16*, 456 (1962).

(10) Finlayson, R. and A. Chrambach, Anal. Biochem. *40*, 296 (1971).

(11) Catsimpoolas, N., Separation Sci. *6*, 435 (1971).

Chapter 6. Preparative Methods

6.1 Theoretical and Practical Aspects of High Resolution Preparative Gel Electrophoresis Using a Continuous Elution System

Stephan Nees

On account of their high separation ability and wide range of application in many separation problems, the techniques of analytical gel electrophoresis rapidly attained a predominant position among available methods of separation. Since the development of these techniques (1) attempts have been made to use them on a preparative scale, in order to separate larger quantities of a mixture of substances into single components. There has been no lack of suggestions for a suitable apparatus (2,4–8), but they have all reflected the considerable technical difficulties associated with the attempt to transfer the analytic conditions to the preparative scale for the following reasons:

1. Preparative electrophoresis systems are more complex than analytical. Factors which in analytical procedures could be neglected can no longer be ignored.
2. The particular behaviour of larger gel masses during extensive periods of electrophoresis and in combination with a continuous elution system first had to be studied and understood in detail on the basis of increasing experience the technique.
3. An elution system had to be developed which met all the requirements and which proved itself a truly productive method, that is: The elution system had to reflect the actual separation capacity of the gel system in separating any two peaks or bands of interest as they migrated out of the gel towards one of the electrodes. The elution device had to permit quantitative recovery of the purified bands, to maintain biological activity, and to avoid excessive dilution.

Before I describe a new preparative electrophoretic system with respect to its construction and applicability, I shall first summarize the theoretical and methodological factors on which the electrophoretic process in large gel masses is based, and which, without exception, have to be taken into account in the design of a suitable apparatus if it is to make possible preparative separations with the same degree of separation efficiency as is known in analysis. I shall not be concerned with special questions which arise with respect to the preparation of, for example, large quantities of nucleic acid or proteins in a form suitable for electrophoretic processing. Questions of this nature have been dealt with in a number of papers (9–11) and in a detailed study, presently being prepared.

Theoretical considerations

1) Hydrostatic pressure between all the buffers and the gel

All the buffer systems used in the apparatus should be in hydrostatic equilibrium with the gel. Should the gel be subjected to a hydrostatic pressure, it can deform and thus distort the band pattern in the gel. Since the hydrostatic equilibrium of the electrode buffers during electrophoresis can be disturbed by electrolytic formation of gas at the electrodes, the gases must be constantly removed from the apparatus as they form.

2) Removal of Joule's heat

The gel represents the greatest electrical resistance in the entire system and is therefore the site of the greatest development of heat when current flows. The resulting heat quantity must be removed from the gel as rapidly and uniformly as possible since many biological macromolecules such as enzymes or cell particles are extremely thermolabile.

The field strength given by $E = J/\kappa$, depends upon κ. In electrolytes (such as for example the gel buffer) κ is dependent on the development of heat because of the increased dissociation of salts on heating. If the gel mass in cross-section is inadequately cooled, and if, for this reason, a temperature gradient develops in the gel, the electrical field is concavely distorted in the direction of migration of the bands. Any distortion of the bands and deviation from the parallelity considerably reduces the possible separation capacity of the gel system used.

3) The shape of the gel

The shape of the gel considerably influences the degree of dissipation of heat from the gel during the process of electrophoresis but also has consequences with respect to the swelling of the gel during electrophoresis, and the efficacy of the elution system. Therefore, the shape of the elution chamber is dictated by the form and cross-section of the gel matrix.

Cylindrically-shaped gels, even when the so-called "cold finger designs" are used, can never be cooled in a perfectly uniform manner since the outer cooling surface is larger than the inner one and the total cooling surface available is relatively small in relation to the gel mass. Gel plates, on the other hand, can be cooled over a wide surface and uniformly from both sides. The efficacy of a cooling system acting on a gel plate in the apparatus described below is represented in Figure 1.

Fig. 1. Separation of human serum proteins in a 7.5 per cent T, 2.63 % C gel using the
Ornstein Davis system. Thirty mg of protein separated at 300 V. Left: from view of gel
slab, Right: cross-section of gel slab. Slab stained with Amido Black.

In the plan and side view one can clearly see the completely parallel alignment
of the bands of the proteins separated in the gel. Under normal conditions of
electrophoresis, gel plates of up to 15 mm in thickness with corresponding
field homogeneity can be used.

4) Material of the gel chamber

The gel is first poured into the separating chamber in the liquid state. In the
chamber, the gel polymerizes and generates heat. The adhesion of the gel matrix
to the walls of the chamber depends upon the polarity of the chamber material
used. Special types of glass (relatively polar) possess a greater affinity for gel
than do apolar plastics material (for example perspex). For this reason, chambers made of such plastics must be rejected since the inadequate adhesion of
the gels to the chamber wall will lead to the production of creepage currents between the gel and the chamber wall which distort the electrical field. Glass chambers have the additional advantage of retaining their plane parallelity even after
being used many times over.

5) Swelling of the gel during electrophoresis

Deformation at the anodal face of gels cast in open tubes is reported in the
literature (2,5,13) and is usually attributed to higher temperature within the
core of the gel (2,13) than in the column walls and buffer reservoir. Kawata (14)
discussed that beyond heat effects there is another major cause of difficulty-

probably ion accumulation with resultant changes in pore size and compression of micelles at the anodal border. Physical supports below the gel would be expecte to further complicate the issuance of substances reaching the gel end. Hjerten (13, 15) therefore abandoned such supports because the pores of a supporting polyethylene disc were found to become clogged in some places as the run progresses. I observed clogging of the gel border while using mechanical supports- probably due to tension and micellar compressions in the gel border which diminis the pore diameter. Porous supports therefore must be rejected. The elution buffer should be in direct contact with the gel, that is, the end of the gel must be directly "washed". This requirement, however, can be satisfactorily met only if a flat gel plate is used. In such a case the adhesive strength of the gel on the walls of the separating chamber is sufficiently large to balance out the swelling of the end of the gel freely dipping into the elution buffer, in such a way that the elution chamber does not become blocked by the bulging gel as occurs in cylindrical gels. However, a slight swelling of the end of the gel is desireable.

6) The Elution System

Shape of the elution chamber: Immediately below the gel, in preparative electro- phoretic devices, there is a conductive chamber having the smallest possible volume, through which a stream of buffer constantly flows. This so-called elution chamber serves to collect the bands migrating out of the gel and to pass them on to a fraction collector. The shape of the elution chamber is determined by the form of the gel: if cylindrical gels are used for electrophoresis, the elution chamber has a circular cross-section, and in the case of gel plates, a rectangular cross- section. For flowtechnical reasons, the flow through elution chambers having a circular cross-section is not uniform, although uniformity of flow *is obtained* in elution chambers with a rectangular cross-section (Figure 2). This also applies to the circular elution chamber model 3 shown in Figure 2 which, at first, seems most attractive with respect to its theoretical consept because of the uniform peripheral supply of elution buffer and the central outflow.

This type of elution chamber, however, will function satisfactorily only if it is possible to screen the gel horizontally from the elution chamber by means of porous supports. As has already been pointed out at 5, however, the gel should be in *direct* contact with the elution buffer. The swelling of cylindrical gels resulting in this case is represented in the side views of elution chambers model 1-3 in Figure 2 and leads to incomplete elution in the peripheral zones of the chamber located directly under the gel. This bulging into the elution chamber causes disturbances in buffer flow with trailing and mixing of zones also in the type elution chamber model 3 of figure 2.

As is shown in the elution diagrams next to the plan and side views of figure 2 (each of the elution chambers compared to one another had the same volume and the same rate of flow of elution buffer), elution chambers having a rectangular

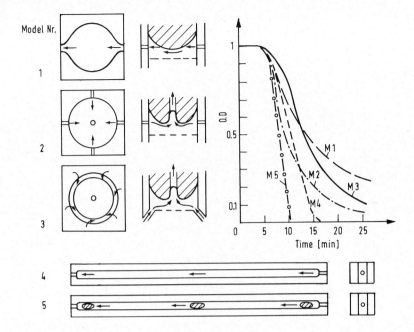

Fig. 2. Comparison of various elution chamber designs and dye elution efficiency curves of each type. All chambers read the same volume and were eluted with the same buffer flow rate of 30 ml/hour in an electric field of 10 V/cm. The graph shows the decrease of the OD of the chamber content during elution.

cross-section were effectively washed by the elution buffer. The maximum elution effect is achieved by introducing air bubbles into the elution buffer at regular intervals which migrate uniformly through the chamber in the elution buffer stream (see model 5, figure 2).

Concentration of the elution buffer: The choice of the correct concentration of elution buffer is an extremely important point in the development of an efficient elution system. The concentration of the elution buffer determines its specific conductivity. This, in its turn, influences, in accordance with the formula $E = J/\kappa$, the electrical field strength in the elution buffer. The main objective in preparative electrophoresis is to ensure that each of the substances arriving at the end of the gel enter as rapidly as possible into the elution buffer which then carries the components out of the apparatus. With respect to the field strength conditions this means that, a higher field strength should be present in the elution buffer than in the gel. Thus, the ratio of buffer concentration of gel buffer and elution buffer is of decisive importance. In principle, three different cases can be described: The concentration (the specific conductivity) of the elution buffer can be greater, identical or lower than that of the gel buffer. The selection of a concentration of the elution buffer greater than that of the gel buffer will result in a stacking of the substance zones arriving at the end of the gel, as a consequence of the fall in the electrical field strength due to the high specific conductance of the elution buffer. Thus, the resolution will be annulled at this point. In addition, the substances will migrate out of the gel only slowly and thus be strongly diluted in the constantly flowing elution buffer. In many biological macromolecules (for example enzymes) this could lead to irreversible denaturing. The osmotic contraction of the end of the gel resulting from the difference in the buffer concentration, may also lead to the obstruction of the pores at the end of the gel by the denatured substances, which would completely negate the elution of separated components.

A decrease of the elution buffer concentration to the concentration of the gel buffer would improve the situation. As a reversal of the consequences of a too high concentration of elution buffer as mentioned above, the choice of a concentration buffer lower than that of the gel buffer leads to optimal separation results. The effect of the osmotic swelling of the end of the gel occurring in this case combines with that of the large field strength gradient in the elution chamber in such a way that the substances arriving at the end of the gel are accelerated virtually exponentially to the free electrophoretic mobility in the elution buffer. The result is an improved separation and a minimal dilution of the substances in the elution buffer. Admittedly, if such an elution system were to be used, the yield of eluted substances would be very low due to substance loss through the membrane of the elution chamber or of absorption and denaturing phenomena at the membrane in consequence of the Bethe-Toropoff (16) effect.

The construction of a second so-called "counter chamber" directly beneath the elution chamber proper, through which a buffer having a concentration of three to four times that of the elution buffer flows, makes it possible to use in practice the previously mentioned low ionic strength elution buffer and still obtain high recovery rates of eluated substance.

Practical consequence: A new apparatus

Figure 3 shows an apparatus[1] developed on the basis of theoretical considerations for the carrying out of preparative and analytical electrophoresis. Technical details can be seen in the legend to the illustration. All the buffer systems of the apparatus can easily be brought into hydrostatic equilibrium. Bubbles of gas generated at the electrodes escape via vent holes in the rear wall. The apparatus makes use of a vertical glass separating chamber having a rectangular cross-section so that the dissipation of Joule heat, the adhesion of the gel, the symmetry of the

---

[1]   manufacturer: Colora-Messtechnik GmbH, 7073 Lorch, GFR

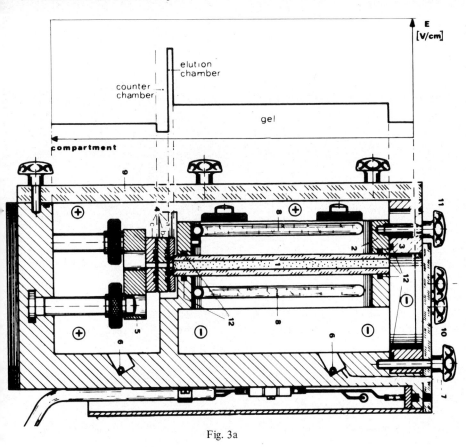

Fig. 3a

Fig. 3. Schematic drawing of the apparatus described in this report. 1 – separation chamber, 2 – vertical frame, 3 – upper frame, 4 – elution system, 4a – elution plate, 4b – counter chamber plate, 4c – base plate, 4d – bore holes, 4e – silicone gasket, 4f – counter buffer tubes, 4g – tubing for air bubble removal, 4h – nylon screws, M – semipermeable membranes, 5 – platform, 6 – electrodes, 7 – vent hole, 8 – indirect cooling system, 9 – front plate, 10 and 11 – cathode and anode buffer surfaces, 12 – gaskets separating the electrode buffers.

gel chamber and a favourable elution cross-section are guaranteed. The elution system is represented in figure 3 in a side view in combination with a schematic representation of the field strength. A perspective representation of the elution system is also shown. In principle it comprises a pile of plates with membranes and ring-seals in between. The apparatus thus functions in accordance with the *counter chamber elution principle* which permits the carrying out of high resolution preparative electrophoresis giving a high yield of eluted substance and associated biological activity. In consequence of the ease of interchangeability and

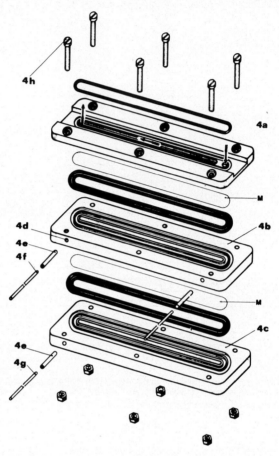

4h
4a
M
4d
4b
4e
4f
M
4c
4e
4g

Fig. 3b

the accessibility of the gel chambers specially dimensioned for various charges and differing purposes (preparative or analytical work), the system can be adapted optimally to any technique of electrophoresis.

Figure 4 shows two characteristic elution diagrams for the separation of human serum proteins and the separation of substances of low molecular weight (dyes). The complete separation of individual serum proteins is, on account of the relative overloading of the albumin band (danger of tailing!), a difficult separation problem and succeeds only when all the theoretical requirements of preparative electrophoresis have been met by the apparatus (see also Fig. 1). All the peaks of the elution diagram shown proved to be pure on reseparation using analytical electrophoresis. The yield of eluted proteins was 98%. Such yields were also foun

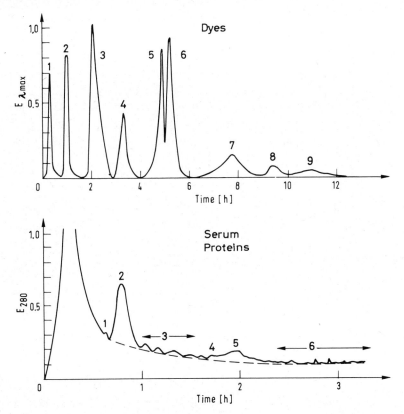

Fig. 4. Elution diagrams of a gelelectrophoretic separation of human serum proteins (disc gel system[1], 300 V, separation gel: 6 cm, load 45 mg) and of low molecular weight dyes (disc gel system[1] but 30 % Acrylamide, 600 V, separation gel: 15 cm). Elution rate in each case: 30 ml/h. 1 Prealbumin, 2 Albumin, 3 Postalbumins, 4 Pretransferrin, 5 Transferrin, 6 Posttransferrins.

for the separation of the individual dyes which have a high electrophoretic mobility. Thus, the efficacy of the counter chamber elution system has so been demonstrated both with respect to the high degree of resolution in the separation process and the recovery of eluted substance components.

With the aid of the apparatus described, analytical and preparative separations can be carried out under optimal and reproducible electrophoretic conditions. The interchangeability and ease of access of the separating chambers permit the carrying out of all common gel electrophoretic techniques. Due to the new type of elution design, this system can also be used for the preparative separation of very small and fast moving substances, a fact which will be of great importance in protein and nucleic acid chemistry: It is now possible to use preparative gel-electrophoresis also for the purification of peptides and nucleotides. Using SDS—Urea—buffer systems (17, 18) it is possible, to purify insoluble material (e.g. core peptides of enzymatic or chemical digestion of proteins).

According to the efficiency of the "indirect cooling system" (see figure 3) in combination with the slab-shape of the gel high retention of biological activity can be obtained. This is especially important when dealing with thermolabile substances (e.g. enzymes, cell particles).

Acknowledgements

For helpful advises during the development of this apparatus are gratefully acknowledged: Drs. L. Bohne; L. Gürtler; K. Röhm; W. Schmidt; W. Zumft.

References

(1)   Davis, B. J., Preprint 'Disc Electrophoresis', Distillation Prod. Div. Eastman Kodak Co., Rochester, N. Y. 1962.
(2)   Yovin, T., M. A. Chrambach, Naughton, Anal. Biochem. 9, 351 (1964).
(3)   Saxena, B. B. and P. H. Henneman, Biochem. J. 100, 711 (1966).
      Abilgaard, U.; Scand. J. Clin. Lab. Invest. 21, 89 (1968).
      Reisfeld, R. A., J. K. Inman, R. G. Mage, and E. Appella, Biochemistry 7, 14 (1968).
      Lewis, K. J., E. V. Cheever, and B. K. Seavey, Anal. Biochem. 24, 162 (1968).
      Gould, J. J., J. C. Pinder, H. R. Matthews, and A. H. Gordon, Anal. Biochem. 29, 1 (1969).
      Simons, K., and A. G. Bearn, Biochem. Biophys. Acta 133, 499 (1967).
(4)   Raymond, S., and E. M. Jordan; Sep. Sci. 1, 95 (1966).
(5)   Schenkein, J., M. Levy, and P. Weis, Anal. Biochem. 25, 387 (1968).
(6)   Duesberg, P. H., and R. R. Rueckert, Anal. Biochem. 11, 342 (1965).
(7)   Brownstone, A. D., Anal. Biochem. 27, 25 (1969).
(8)   Smith, J. K., and D. W. Moss, Anal. Biochem. 25, 500 (1968).
(9)   Bishop, D. H. L., and J. R. Claybrook, S. Spiegelmann, J. Mol. Biol. 26, 373 (1967).
(10) Weber, K., and M. Osborn, J. Biol. Chem. 244, 4406 (1969).
(11) Nees, S., and K. Richter, Instruction manual 'UltraPhor' (1972); Colora Messtechnik GmbH 7073 Lorch (GFR), p 42.
(12) Skyring, G. W., R. W. Miller, and V. Purkayastha, Anal. Biochem. 36, 511 (1970).
(13) Hjertén, S., S. Jerstedt, and A. Tiselius, Anal. Biochem. 27, 108 (1969).
(14) Kawata, H., M. W. Chase, R. Elyjiw, and E. Machek, Anal. Biochem. 39, 93 (1971).
(15) Hjertén, S., S. Jerstedt, and A. Tiselius, Anal. Biochem. 11, 211 (1965).
(16) Svenssen, H., Advanc. Protein. Chem. 4, 251 (1948).
      Manegold, E. and K. Kalauch, Kolloid-Z. 86, 313 (1939).
      Hochstrasser, H., L. T. Skeggs, K. E. Lentz, and J. R. Kahn, Anal. Biochem. 6, 13 (1963).
(17) Swank, R. T., and K. D. Munkres, Anal. Biochem. 39, 462 (1971).
(18) Parish, C. R., and J. J. Marchalonis, Anal. Biochem. 34, 436 (1970).

6.2 Preparative Gel Isoelectric Focusing

D. Graesslin and H. C. Weise

Introduction

Preparative isoelectric focusing (IEF) is generally performed in columns in a sucrose density gradient. The effectiveness of this technique, however is restricted by the necessity of draining the column, which leads to diffusion and partially remixing of primarily clearly separated adjacent protein zones. Further drawbacks are the long focusing time and the removal of sucrose. An attempt to overcome some of the difficulties of IEF in liquid media was made by Radola et al. (1), who used thin layers of Sephadex gel for preparative purposes. Nevertheless, the chance of isolating homogeneous protein fractions by this technique is reduced and limited as the localisation of protein bands can be performed only indirectly by staining a print of paper taken from the gel.

In order to utilize the excellent resolving power of thin layer IEF in polyacrylamide gel, we scaled up the analytical method (2, 3, 4) to preparative purposes. Here, a rapid localisation of protein bands is possible by a selective Coomassie Brillant Blue staining or by direct u. v. evaluation. The employed technique will be demonstrated in this paper on the examples of colored proteins and human chorionic gonadotropin (HCG), a multicomponent complex of glycoprotein isohormones, and the results obtained are compared to the column procedure.

Materials and Apparatus

Acrylamide, N, N' — methylenebisacrylamide, both of analytical grade, sucrose of ultrapure quality, and Coomassie Brillant Blue R 250 were purchased from Serva, Heidelberg, Germany. Myoglobin from horse and sperm whale were supplied by Schuchardt, München, Germany. Crude Human Chorionic Gonadotropin (HCG) obtained from Organon, Oss, Holland, was purified as described earlier (5). Ampholine (LKB–Producter, Stockholm), pH 3–10 and 3–6 was used as commercially available. Plates of high purity transparent quartz glass, 260 × 170 × 5 mm were obtained from Westdeutsche Glasschmelze, Geesthacht, Germany. A well closed plastic box for IEF runs in a refrigerator of size 32 × 24 × 8,5 cm and equipped with carbon electrodes, 20 mm ϕ and 200 mm long (EK 506, Ringsdorf-Werke, Godesberg, Germany).

For gel slabs with increased thickness sufficient cooling is achieved with the Thin Layer IEF Double Chamber from Desaga, Heidelberg, Germany. Carbon electrodes are used instead of CAF contact bridges.

Densitometric u. v. measurements were done with the Zeiss spectrophotometer with thin layer equipment (Zeiss, Oberkochen, Germany).

Methods

Preparative flat bed IEF is performed in gels coated to glass plates of 26 × 17 cm in size. The polyacrylamide gel layers have a thickness between 2 and 4 mm, corresponding to the protein load needed, and contain 1 % carrier ampholytes. 50 ml solution for the gel of 2 mm thick slabs is prepared by mixing 7.5 ml solution I (100 g of acrylamide and 2.7 g of N,N' — methylene bisacrylamide in 300 ml distilled water), 2,5 ml solution II (3 mg riboflavin in 100 ml distilled water), 0.5 ml Ampholine (40 %) of pH 3—10 or 3—6, and 40 ml dist. water. The mixture is pipetted between 2 clamped glass plates, held 2 mm apart by a silicone tube. Siliconizing of one of the plates is unnecessary. Vertical alignement is maintained. For Photopolymerisation the gel is allowed to develop for 90 min. in front of two 6 watt day light fluorescent tubes arranged at a distance of 15 cm. After polymerisation clamps and silicon tube are removed and at one edge a razor blade is drawn between gel and one glass plate. This plate is then carefully lifted off with a knife. Unpolymerized components are removed by washing the gel slab in distilled water for about 24 hours at 4° C under gentle shaking. The washings are renewed twice. The gel remains adhered to the plate. After about two hours at room temperature, when the gel surface is nearly dry, 2,0 ml of Ampholine (40 %) are equally spread over the gel by means of a glass rod. Exact horizontal alignment is maintained. When the ampholytes are soaked in and the gel surface is nearly dry again (1—2 hours), the slab is ready for use after a short time of precooling.

The sample can be applied in several ways:

10—20 mg of proteins, dissolved in 0.3 ml 1 % Ampholine, are soaked in a 1 cm broad and 10 cm long paper strip (Whatman No. 1) and are put on the gel surface. Alternatively, 0.4 ml of a 5 % solution of proteins in 1 % Ampholine may be layered on the gel surface in streak form with a micropipette.

When gel slabs of 4 mm thickness are used, preferably the sample is applied by soaking 0.6 ml of a 5 % protein solution into a 100 mm long, 4 mm broad and 1 mm thick paper strip (Schleicher & Schüll, No. 2668), which is inserted with a forceps into an adequate slit of the gel.

Runs in 2 mm layers may be performed in a humid plastic box in the refrigerator at 2° C, whereas for 3 and 4 mm gels a more effective metal cooling system with circulating water-methanol of 4° C is required.

Carbon electrodes are put directly on the gel at a distance of 100 mm. The protein samples can be applied immediately or about 30 min. after starting of

the run. The starting voltage of 100 V is increased to 3–400 V after 20 min.
Total focusing time requires between 5 and 8 hours.

For localisation of the protein bands after the run, 25 mm apart from both
edges of the slab 5 mm discs are punched along the gel with a cork borer at
equidistant intervals. After cutting clean through the diameter of the punches,
the two parallel strips are loosened from the plate and are stained for 20 min.
in a 0,1 % solution of Coomassie Brillant Blue R 250 in water: ethanol: acetic
acid (50 : 45 : 5), with mechanical agitation and subsequently partially destained
in water: ethanol: acetic acid (55 : 40 : 5) for about 15 min., under gentle
shaking (4). The band positions in the unstained gel are identified by replacing
the stained gel strips on their former places. The localisation procedure is shown
in figure 1.

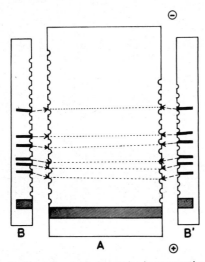

Fig. 1. Technique of localisation of separated proteins in preparative thin layer IEF.
A, unstained middle part of the gel slab. B and B', Coomassie Blue stained gel strips. The
places of corresponding protein bands are indicated by marker holes.

In spite of some shrinking of the gel during the staining procedure the semicircles
of the punched gel discs allow to mark reliably the corresponding positions of
identical protein bands. Now these bands in the middle part of the gel are cut
out by means of a razor blade and the proteins are eluted by about 10 ml of
distilled water or buffer at 4° C with gentle agitation. The Ampholine is removed
either by dialysis or by gel filtration on Sephadex G 25 in a 9 × 200 mm column.

If the gel is coated to a plate of quartz glass, quick localisation of the protein
bands is also possible by direct u. v. scanning with a Zeiss spectrophotometer at
280 nm.

Results

In order to demonstrate the excellent resolving conditions, even in a 4 mm thick polyacrylamide gel layer, 30 mg of a mixture of horse myoglobin, sperm whale myoglobin and bromophenolblue albumin were separated in the pH range 3–10. Figure 2 shows the unstained pattern at the end of the run. The myoglobins appear as sharp, well resolved parallel bands in the pI range 6.9 to 8.3, as well as colored albumin at pH 4.9. The protein sample had been applied near the anode.

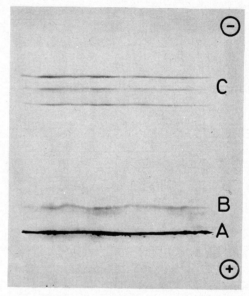

Fig. 2. Unstained band pattern of 30 mg of a mixture of horse, whale myoglobin and Bromophenolblue albumin after isoelectric focusing in 4 mm polyacrylamide gel layers (pH 3–10, 35 V/cm, 7 hours). A, sample application, B, colored albumin, C, myoglobins.

When 15 mg of highly purified HCG was fractionated by preparative IEF on 2 mm thick gel slabs, it separated into 6 distinct components, five of which were isolated, yielding each between 1.5 and 2.5 mg biological active proteins. Total protein recovery was 65 %. Figure 3 demonstrates the homogeneity of these isohormone preparations when aliquots were refocused in analytical scale in the same pH range, 3–6. The isoelectric points are given in figure 3. The biological and immunological properties have been reported earlier (5).

For comparison studies 20 mg of highly purified HCG were subjected to column IEF in a sucrose density gradient. The pH range was 3–6, the column capacity

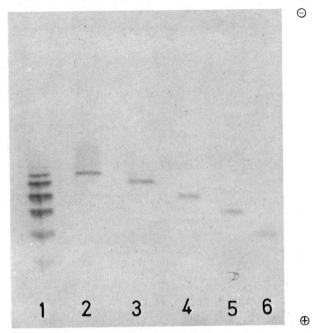

Fig. 3. Analytical gel IEF pattern of aliquots of 5 HCG fractions (positions 2–6), isolated
by preparative gel IEF. Pos. 1: starting material. pH 3–6 Ampholine. Staining with
Coomassie Brillant blue.

110 ml. Again five corresponding factions could be isolated. Refocusing in the
analytical gel technique, however revealed, that each fraction contained one
major band still overlapped by one or two minor adjacent zones. This indicated
clearly, comparing the "column" and the "thin layer" technique, that only the
latter yielded homogeneous fractions.

The second possibility of quick localisation of protein bands is given in
figure 4a. It shows the u.v. pattern of unstaines HCG components after separation in 2 mm gel layers, pH 3–6, obtained by direct fotometrical evaluation
at 280 nm. For comparison, the elution pattern of the HCG fractions after
separation in column IEF is seen in figure 4b. The overlapping peaks indicate,
that a complete resolution of protein components could not be achieved.

Discussion

Up to now preparative IEF was described in columns in a sucrose density
gradient and more recently on glass plates coated with layers of Sephadex gel
(1) and in tubes of polyacrylamide gel (7). It is known, however that draining

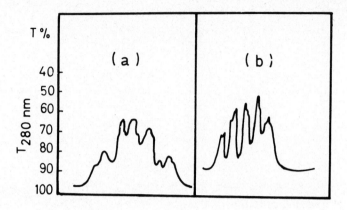

Fig. 4. Scan at 280 nm of 20 mg highly purified HCG after column IEF in sucrose density gradient (a), and 15 mg highly purified HCG, after preparative thin layer IEF in poly-acrylamide on quartz plates (b).

of the column leads to diffusion, turbulances and viscous flow effects. Thus isolation of homogeneous protein fractions often is prevented. On the other hand the effectiveness of the thin layer technique of Radola is restricted by the necessity of indirect localisation of separated zones, by staining a print of paper taken from the gel. This procedure including fixing, washing, staining and destaining is very time-consuming. In addition, the degree of wetness decidedly affects the resolving power of this technique.

Preparative application of IEF in polyacrylamide gels was firstly conducted in tubes by Finlayson and Chrambach (7). The load capacity was reported to be 2 mg of protein per band per single gel. However the isolation of native proteins thus was not possible as proteins were fixed with trichloroacetic acid for visualisation and extracted from the gels with 0.1 M formic acid.

Most of these drawbacks could be overcome by scaling up the analytical flat bed IEF in polyacrylamide gel to preparative purposes, as described in this paper.

The excellent resolving power of analytical gel IEF, as described by Awdeh (2) is fully retained. Direct and quick detection of the band positions is possible in two ways: by using a selective staining with Coomassie Brillant Blue or by employing photometrical u. v. evaluation of proteins in gels coated to quartz plates. We have successfully performed runs in polyacrylamide gel with different thickness. Up to 40 mg of proteins can be applied to 4 mm layers. Two satis-factory loading procedures have been used. The technique and apparatus is simple.

Using this technique we succeeded in separating and isolating five HCG iso-hormones, which are electrophoretically and immunologically homogeneous and possessed high biological activity each. On refocusing the components appeared as single bands.

The superiority of the polyacrylamide gel technique clearly can be seen comparing the two refocusing patterns of HCG fractions obtained by the thin layer and the column method, respectively. Preparative gel IEF yielded homogeneous protein fractions, while the column technique resulted in overlapping of adjacent bands. (figures 4a, b). A further methodological comparison was possible since Brossmer et al. (6) also reported separation of highly purified HCG into several biological active fractions. The authors used the preparative IEF on thin layers of Sephadex G 75, as described by Radola (1). The isolated fractions displayed homogeneity when refocused under the same conditions, however refocusing on thin layers of polyacrylamide gel showed a further splitting into several bands.

Summary

A preparative method for separating proteins by isoelectric focusing (PIEF) in polyacrylamide gel is described. The high resolving power of analytical IEF is fully retained. PIEF is conducted in gel slabs up to 4 mm with a load capacity up to 40 mg. Using this technique the isolation of 5 biologically active HCG isohormones was possible.

Acknowledgements

This study was supported by a grant from DFG — Sonderforschungsbereich 34 — Endokrinologie. We thank Mrs. A. Trautwein for excellent technical assistance.

References

(1)   Radola, B. J., Biochim. Biophys. Acta *194*, 335 (1969).
(2)   Awdeh, Z. L., A. R. Williamson, and B. D. Askonas, Nature (London) *219*, 66 (1968).
(3)   Leaback, D. H., and A. C. Rutter, Biochem, Biophys. Res. Commun. *32*, 447 (1968).
(4)   Graesslin, D., A. Trautwein, and G. Bettendorf, J. Chromatogr. *63*, 475 (1971).
(5)   Graesslin, D., H.-Chr. Weise, and P. J. Czygan, Febs Letters *20*, 87 (1972).
(6)   Brossmer, R., W. E. Merz, and U. Hilgenfeldt, Febs Letters *18*, 112 (1971).
(7)   Finlayson, G. R., and A. Chrambach, Anal. Biochem. *40*, 292 (1971).

Chapter 7. Micro Methods

7.1 Micro Disc Electrophoresis in Capillary Columns

Ulrich Grossbach

The method of disc electrophoresis on polyacrylamide gels (Ornstein 1964, Davis 1964) has already been adapted to the separation of proteins in the millimicrogram range (Grossbach 1965, 1969, 1971). Microelectric focusing of protein mixtures in capillary gels on the same level has also been described (Grossbach 1972).

For a successful approach to a wide variety of problems in cell biology, it would, however, be desirable to further increase the sensitivity of the technique. Edström's method of microelectrophoresis on silk threads, which allows the separation and determination of nucleic acid bases in the micromicrogram range (Edström 1956, 1960a, 1964), has been extremely useful for the analysis of single cells and small groups of cell organelles (Daneholt and Edström 1969, Edström 1960b, Edström and Beermann 1962, Edström and Gall 1963, Edström et al. 1961, Egyhazi and Hydén 1961, Hydén and Egyhazi 1962). A method of similar sensitivity for the separation of proteins may become comparably advantageous.

That it is possible to perform gel electrophoresis on the micromicrogram scale was very elegantly demonstrated by Matioli and Niewisch (1965), who separated the hemoglobins of single erythrocytes which had been polymerized into a fiber of polyacrylamide. In the present communication, a generally applicable method is described for the separation of proteins in the micro-microgram (picogram) range.

Methods

Electrophoresis is performed in gels of $50\ \mu$ and $100\ \mu$ diameter. Quartz-glass capillaries with such internal diameters can be obtained from Heraeus Quarz-schmelze (Hanau, Germany). Especially uniform segments are selected by means of an eyepiece micrometer and cut to lengths of 25 mm. The total volume of the capillary columns is thus $0.05\ \mu l$ and $0.2\ \mu l$, respectively.

An 8 mm long silicone rubber stopper with suitable bore is fastened onto the capillary. Both sides of the stopper fit into a rubber ring which is fastened to the instrument holder of a micromanipulator. The capillary is thus attached to the manipulator in a perpendicular position, with both ends freely accessible. This arrangement allows controlled movements of the capillary column in relation to a micropipette which is mounted on a second micromanipulator (figure 1).

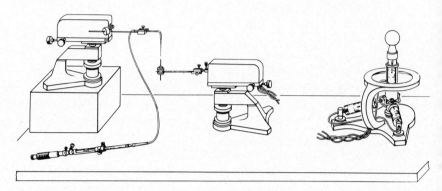

Fig. 1. Micromanipulator set-up for the preparation of capillary gel columns. Details are given in the text.

Because of the pneumatic transmission of movements in all three dimensions, the de Fonbrune (1949) micromanipulator is especially suited for use in micro-electrophoresis. With the instrument adjusted to the smallest reduction of the hands movements, the capillary can be moved maximally 3 mm in each direction. Movements over a larger range are exerted by the rack-and-pinion device of the instrument.

Micropipettes which can be inserted into the capillary columns are manufactured from capillaries of soft glass, having an internal diameter of about 1 mm. These are pulled out in a small gas flame. Continuously tapered pipette tips with an orifice of about 10 μ diameter are then pulled over a 1 mm long microflame, which is produced by attaching a suitable glass capillary to the gas outlet of a Bunsen burner.

The best pipettes are selected under a microscope, and are sealed with paraffin into a piece of 3 mm glass tubing which has been bent to form a right angle. The glass tubing is attached to the instrument holder of a micromanipulator in a way which orients the pipette perpendicularly (figure 1). It is then connected by a piece of silicone rubber tubing to an airfilled syringe. The piston of the syringe can be operated free-hand or with a micrometer screw (figure 1) for the delivery of very small volumes of solution. Capillary and pipette manipulations were observed with a stereomicroscope, with illumination from behind.

For the preparation of polyacrylamide gels, the solutions and buffers of the standard system for discontinuous electrophoresis (Davis 1964) are used. Capillary columns are filled with small-pore gel solution to a suitable height by capillary action. Water is layered over the gel solution by means of the micropipette, which has been detached from the syringe and filled to a point at which teh capillary action balances the pressure of the water column. When the pipette tip is brought in contact with water, the level of the water in the pipette does not change; however, when the pipette tip touches the upper meniscus of the gel solution in the capillary, water is very slowly delivered as a result of the higher viscosity of the solution. By this means, careful over-layering of the gel solution with water, which is important for the formation of a uniform gel surface, is easily achieved.

After polymerization, the capillary column is inverted, and the lower end of the gel is covered with electrode buffer by means of the pipette. Evaporation is prevented by protecting this end with a small amount of glycerol. The column is then returned to its original position on the instrument holder, and the water on the gel surface is removed with the pipette. The gel surface is rinsed several times with stacking-gel solution to remove traces of water. It is finally overlayered with a volume of this solution determined by the volume of the protein sample, and is photopolymerized under a thin layer of water.

The protein sample is dissolved in stacking-gel solution containing 20% sucrose, applied on top of the stacking-gel, and carefully overlayered with electrode buffer. The capillary is then detached from the micromanipulator and, by means of the rubber stopper, connected to a piece of 5 mm glass tubing which can be attached to a conventional apparatus for gel electrophoresis (figure 2).

Electrophoresis is performed at a constant voltage of 10 volts for the first 2 min, and subsequently at 100 volts for 5—10 min. Because of the short time needed for the separation, electrophoresis at constant voltage was considered to be not inferior to constant current conditions.

Immediately after the end of the run, the gels are pushed out of the capillaries with a piece of metal wire. Rigid tungsten wire of 50 μ diameter can be obtained from Callite Co. and stainless-steel wire of 100 μ diameter from Nirosta Westfalia. It is important that the wires fit exactly into the capillaries, as the gels will otherwise be destroyed.

Fixing and staining is performed on a siliconized depression-slide. The gels are fixed for 1 h in 20% trichloroacetic acid, and stained overnight in 0.1% coomassie brilliant blue (Meyer and Lamberts 1965). Destaining is achieved within a few minutes, first in methanol — 7% acetic acid (1:9), and then in a solution of ethanol, acetic acid, glycerol, and water (5:2:5:13).

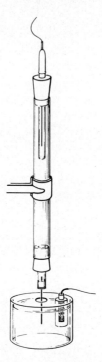

Fig. 2. Arrangement for micro disc-electrophoresis in capillaries.

The gels can be photographed on Polaroid film 55P/N with a Polaroid MP-3
camera and a Rodenstock Eurygon lens (1:4, f = 35 mm). They are stored in
1 μl capillary pipettes (Drummond microcaps), sealed with paraffin at both
ends.

Results

Figures 3 and 4 show the separation of the monomer and dimer fractions of
bovine serum albumin (crystalline, Armour). 2.7 mμg of albumin were applied
to a gel of 100 μ diameter (figure 3). The sample contained 5.0% dimer, as
determined by densitometry of gels with a diameter of 5 mm run in parallel.
The dimer band in figure 3 therefore represents 140 $\mu\mu$g of protein.

Figure 4 shows the separation of 340 $\mu\mu$g of albumin from the same sample on
a gel of 50 μ diameter. In this case, the dimer band represents 17 $\mu\mu$g, and the
monomer zone 323 $\mu\mu$g of protein.

In both separations, the amount of albumin monomer applied was too high,
giving rise to a broad zone of protein, while the dimer fraction formed an

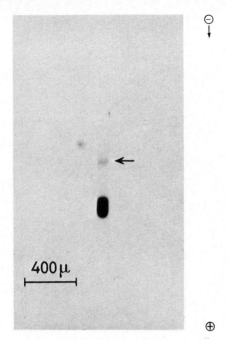

Fig. 3. Micro disc-electrophoresis of 2.7 mμg serum albumin on 15% polyacrylamide in a capillary of 100 μ diameter. The dimer band (arrow) contains 5.0% i. e. 140 $\mu\mu$g of protein.

appropriate band. As was found by electrophoresis on 5 mm wide gels, the albumin sample also contained about 0.5% of an oligomer, but this component was not detected by microelectrophoresis under the conditions applied.

It would seem, therefore, that micro disc-electrophoresis in capillaries of 50 μ diameter lends itself to the separation of proteins in the range of 10^{-11} to 10^{-10} g, while capillaries of 100 μ diameter are appropriate for the analysis of 10^{-10} to 10^{-9} g of protein. The technique is thus as sensitive as Edström's micromethod (microphoresis) for nucleic acid bases. It may therefore become a similarly useful tool in cell biology, e. g. in studies of cell differentiation.

Acknowledgements

My sincere thanks are due to Professor Jan-Erik Edström, who taught me the technique of micromanipulation. I am indebted to Professor Wolfgang Beermann for his kind support. I also thank Mr. E. Freiberg for drawing the figures,

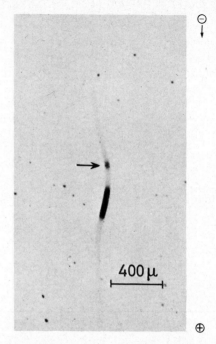

Fig. 4. Micro disc-electrophoresis of 340 μμg serum albumin on 15% polyacrylamide in a capillary of 50 μ diameter. The dimer band (arrow) contains 5.0% i. e. 17 μμg of protein.

and Miss E. Kasch and Mr. J. Brand for their excellent assistance. The work was supported by a grant from the Deutsche Forschungsgemeinschaft, which is gratefully acknowledged.

References

Davis, B. J., Ann. N. Y. Acad. Sci. *121*, 404 (1964).
Daneholt, B., and J.-E. Edström, J. Cell Biol. *41*, 620 (1969).
Edström, J.-E., Biochim. biophys. Acta *22*, 378 (1956).
Edström, J.-E., J. Biophys. Biochem. Cytol. *8*, 39 (1960a).
Edström, J.-E., J. Biophys. Biochem. Cytol. *8*, 47 (1960b).
Edström, J.-E. In: Methods in cell physiology (ed. D. M. Prescott), vol. 1, p. 417, New York (Academic Press) 1964.
Edström, J.-E. and W. Beermann, J. Cell Biol. *14*, 371 (1962).
Edström, J.-E. and J. G. Gall, J. Cell Biol. *19*, 279 (1964).
Edström, J.-E., W. Grampp and N. Schor, J. Biophys. Biochem. Cytol. *11*, 549 (1961).
Egyhazi, E. and H. Hydén, J. Biophys. Biochem. Cytol. *10*, 403 (1961).

de Fonbrune, P., Technique de Micromanipulation, Monographies de l'Institut Pasteur, Paris (Masson) 1949.

Grossbach, U., Biochim. biophys. Acta *107*, 180 (1965).

Grossbach, U., Chromosoma *28*, 136 (1969).

Grossbach, U., Habilitationsschrift, Universität Stuttgart-Hohenheim (1971).

Grossbach, U., Biochem. biophys. Res. Commun. *49*, 667 (1972).

Hydén, H. and E. Egyhazi, J. Cell Biol. *15*, 37 (1962).

Matioli, G. T. and H. B. Niewisch, Science *150*, 1824 (1965).

Meyer, T. S. and B. L. Lamberts, Biochim. biophys. Acta *107*, 144 (1965).

Ornstein, L., Ann. N. Y. Acad. Sci. *121*, 321 (1964).

7.2 Micro Electrophoresis and Its Applications

V. Neuhoff

The technique of micro electrophoresis with polyacrylamide gels has been shown to be as useful as the normal scale procedure and to have the advantage to be much more sensitive and much less time consuming. 0.1 to 0.5 μg of protein can be subjected to electrophoresis in 5 μl capillaries, that means 5—10 000 runs can be performed with 1 ml containing 1 mg of protein. The lower limit for a single band in a 5 μl gel with 450 μm diameter is 10^{-9} g of a protein stainable with amido black 10 B. The electrophoresis time is dependent on the type of protein to be fractionated, and generally takes between 20 and 60 min. Staining with amido black requires only 5 min and the decolorizing process about 30 min. An electrophoretic destaining procedure is not necessary.

Enzymatic activity can be recovered from protein bands as has been shown with DNA- and RNA poymerase from E. coli (Neuhoff und Lezius, 1967, 1968; Neuhoff, Schill und Sternbach 1968, 1969a, b, c 1970). Individual proteins separated during electrophoresis can be re-isolated (Neuhoff und Schill, 1968). The electrophoretic elution method can be applied to all small proteins, like albumin. To elute large protein particles, like RNA-plymerase, the isolated gel discs containing the protein fraction of interest must be frozen at -170^0 C in water free glycerol, slowly thawed and then eluted electrophoretically (Neuhoff, Schill und Sternbach, 1969b). The removed protein can then be used for enzyme activity tests, microdialysis (Neuhoff und Kiehl, 1969), micro analysis of amino acid composition (Neuhoff, Weise, Sternbach, 1970) or micro immunoprecipitation (Neuhoff und Schill, 1968).

Enzyme kinetics of dehydrogenases can also be performed in microgels (Cremer, Dames, Neuhoff 1972). Generally 0.3—1.5 μl of diluted sample containing the dehydrogenase to be tested is subjected to electrophoresis. The optimal concentration has to be determined separately for each sample, the final dilution of the extracts is made just prior to electrophoresis. After electrophoresis the gels were transferred into a tetrazolium mixture containing NAD or NADP and the proper substrate depending on the dehydrogenase the activity is to be measured. The blue formazan which is precipitated as the final product of the coupling reactions at the enzymatically active side in the gel can then be recorded densitometrically and the peak areas measured by planimetry. In this case one can demonstrate that the peak area increases linearly with incubation time. Even when only 0.9 μl of a glucose-6-phosphate dehydro genase solution corresponding to 0.64 pg G6PD is fractionated on a 5 μl gel, a linear correlation (correlation coefficient r = 0.996) beween peak area and time of incubation could be followed for up to a period of 80min. If the reaction rate or the linear increment from experiments with different amounts of G6PD is plotted against the amount of enzyme subjected to electrophoresis one obtains a regression line which can be used as a calibration plot for the quantitative determination of enzyme concentration. The method is sufficiently sensitive to allow quantitative determinations with extracts from only a few μg of fresh tissue. As an example, a single mouse ovum contains sufficient G6PD to allow several assays of the G6PD activity (Dames, Cremer, Neuhoff 1972). Furthermore the method allows G6PD-variants to be analysed in a single step procedure as was shown for the separation of G6PD variants of man and rat and man and mouse. For a further characterization of a G6PD variant it is possible to measure its binding constants after fractionation on microgels (Cremer, Dames, Neuhoff 1972). It is therefore possible to compare a microgel to a "microcuvette" with a volume from 0.003 to 0.015 μl, since a single protein band in a 5 μl gel extendes from approximately 20 to 100 μm.

The micromethod for isoelectric focussing (Quentin and Neuhoff, 1972) requires only a few μg wet weight of tissue to allow determination of LDH isoenzymes from various tissues and various regions of the rabbit brain. IEF of the clear supernatants is caried out at a constant voltage of 200 V at room temperature in 5 μl capillaries filled with a proper gel ampholine mixture (acrylamide 6.4%, crosslinking 2.5%, ampholine 3.8%, saccharose 12%). After about 50 min. the electrophoresis is terminated and the gels are transferred to the tetrazolium incubation mixture, containing lactate as the substrate, for measuring the enzymatically active proteins. Incubation is carried out in the dark at 37^0 C and terminated by transferring the gels into 7.5% acetic acid. Kinetic measurements of the different isoenzymes are also possible by recording the enzyme activity at defined time intervals. It was shown that each of 15 analyzed brain regions has a specific composition of LDH isoenzymes (Quentin and Neuhoff, 1972).

Micro disc electrophoresis has been combined with antigen-antibody crossed electrophoresis in vertical agarose gel (Dames, Maurer, Neuhoff, 1972). The vertical arrangement of the immunoelectrophoresis in closed glass cells guarantees an excellent contact between agarose gel, polyacrylamide gel and the electrode buffers, and apparently reduces endosmotic flow of liquid. Less than 1 μg of antigen protein and about 20 μl of antiserum are needed for one crossed electrophoresis.

For technical details see Neuhoff (1973).

References

Cremer, Th., W. Dames and V. Neuhoff, Hoppe-Seyler's Z. Physiol. Chem. *353*, 1317 (1972).

Dames, W., Th. Cremer and V. Neuhoff, Die Naturwissenschaften *59*, 126 (1972).

Dames, W., H. R. Maurer and V. Neuhoff, Hoppe-Seyler's Z. Physiol. Chem. *353*, 554 (1972).

Neuhoff, V., Micromethods in Molecular Biology, in: Molecular Biology, Biochemistry and Biophysics, Eds. A. Kleinzeller, G. F. Springer and H. G. Wittmann, Springer Verlag, Berlin, Heidelberg, New York, 1973, Vol. 14.

Neuhoff, V., F. v. d. Haar, E. Schlimme and M. Weise, Hoppe Seyler's Z. Physiol. Chem. *350*, 121 (1969).

Neuhoff, V. und F. Kiehl, Arzneim. Forsch. *19*, 1898 (1969).

Neuhoff, V. und A. Lezius, Hoppe-Seyler's Z. Physiol. Chem. *348*, 1239 (1967).

Neuhoff, V. und A. Lezius, Z. Naturforsch. *23b*, 812 (1968).

Neuhoff, V., W.-B. Schill und H. Sternbach, Hoppe Seyler's Z. Physiol. Chem. *349*, 1126 (1968).

Neuhoff, V., W.-B. Schill und H. Sternbach, Arzneim.-Forsch. *19*, 336 (1969).

Neuhoff, V., W.-B. Schill und H. Sternbach, Hoppe-Seyler's Z. Physiol. Chem. *350*, 335 (1969).

Neuhoff, V., W.-B. Schill und H. Sternbach, Hoppe-Seyler's Z. Physiol. Chem. *350*, 767 (1969).

Neuhoff, V., W.-B. Schill und H. Sternbach, Biochem. J. *117*, 623 (1970).

Neuhoff, V., M. Weise and H. Sternbach, Hoppe-Seyler's Z. Physiol. Chem. *351*, 1395 (1970).

Quentin, C.-D. and V. Neuhoff, International J. Neuroscience *4*, 17 (1972).

7.3 A Method for Continuous Gradient Gel Electrophoresis in Capillary Tubes

R. Rüchel

Continuous acrylamide gel gradients can be readily produced without any gradient forming apparatus simply by using capillary attraction. If the stock solutions of the gel are divided in such a way, that one compartment contains

all components without persulfate and the other compartment contains only
the persulfate, a continuous gel gradient is achieved by sucking first the
persulfate solution (sol. 1) and then the acrylamide, buffer and accelerator
solution into a capillary (sol. 2 + 3). The lancet-like penetration of the lower
fraction into the upper persulfate solution (figure 1) will vary depending upon

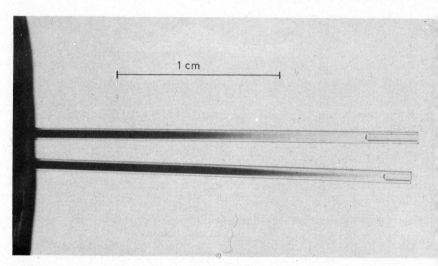

Fig. 1. 5 μl capillaries filled with gel mixture. Immediately after filling (right side) and
after 5 minuts of diffusion (left side).

the diameter of the capillary, the viscosity, the density and the surface
tension of the solutions. Since the shape of the gradient in its vertical
direction is achieved at the moment of capillary flow, the distribution in the
horizontal direction must be achieved by diffusion and requires about 10 min.
in a 5 μl capillary, therefore, the polymerization should not begin during this
time period. The persulfate gradient proceeds from the top to the lower end.
Using a stock solution of the 20%-gel and 5 μl capillaries, gel gradients of
15 mm length can be obtained. Above and below the gel remains an excess
of persulfate and acrylamide solution, respectively.

Gel gradients with gamma globulin incorporated into the acrylamide fraction
show, after staining and densitometry, a nearly linear slope.

Figure 2 shows the relation between the log of the molecular weight against
the relative migration distance of human albumin and its polymers. This
indicates the linearity of the gradient (1).

The upper part of the gradient has about 1% acrylamide concentration (large
pore size) and the lower end of the gradient is about 40% acrylamide, it

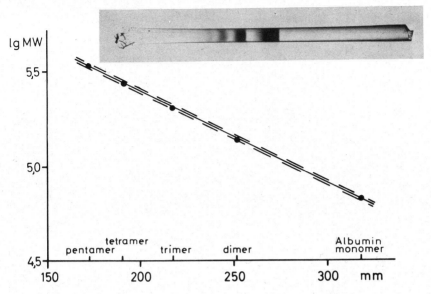

Fig. 2. Separation of human albumin and polymers on a 5 μl gel gradient. Plot of the log MW against the linear relative migration distance obtained by densitometry.

appears knob-like. The stability of the lower end is used to push the gel out of the capillary with a steel wire. When detergents like Triton are present in the gels, water pressure can be used to extrude the gel. However, when SDS is used in the fractionation system, Triton cannot be included. Both continuous and discontinuous buffer systems can be used with SDS. Figure 3 shows the separation pattern of myelin proteins from rat brain following discontinuous SDS-electrophoresis in a 5 μl gradient gel. (Extraction of proteins by Dr. T. Waehneldt, Max-Planck-Institut für experimentelle Medizin, Göttingen (2).) We have found that the most useful buffer system was a "discontinuous voltage gradient" system as described by Allen et al. (3). The "cap gel" of the original system can be either omitted or substituted by dilluted gel buffer.

To obtain high concentration of the sample ions, the ionic strength of the gel buffer should be as high as possible. Although the amperage is high, deformation of the bands, normally associated with heating, does not occur, this is due to the favourable relationship between circumference and surface area of the protein disk which allows constant dissipation of heat.

At 60 volts serum protein fractionation takes place within 20 minutes. Reproducibility of the method is shown in figure 4 by densitometry of six gels, produced under the same conditions.

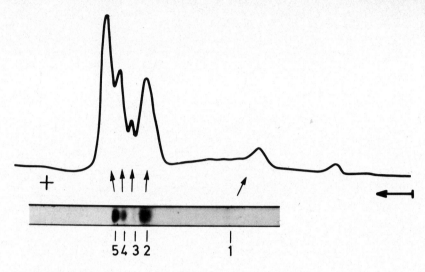

Fig. 3. Separation of myelin proteins from rat brain on a 5 μl gel gradient, discontinuous buffer system with 0,1 % SDS (2) 1: Wolfgram protein, 2: Proteolipid, 3: Agrawal protein, 4 and 5: Basic proteins.

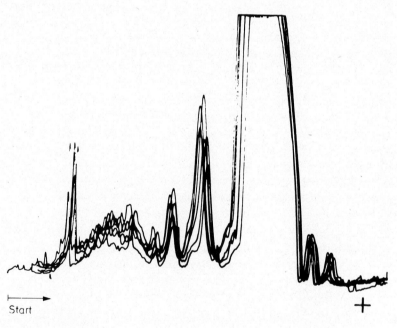

Fig. 4. Separation of human serum proteins on six different gradient gels. Electrophoresis and densitometry under identical conditions.

Protein separation in a gel gradient is mainly dependent on the molecular size as shown by electrophoresis of LDH isoenzymes from rat brain (figure 5). Figure 5 a shows a separation on a gel gradient. Figure 5 b shows a separation on a 10% acrylamide gel (buffers and running time as in figure 5 a, Figure 5 c shows a separation by isoelectrofocusing. All separations were carried out in 5 μl capillary gels. The enzymes are stained by incubation in the tetrazolium system (4) and densitometry is performed under the same conditions. (Extraction of LDH and Isoelectrofocusing by G. Bispink, Max-Planck-Institut für experimentelle Medizin, Göttingen.) Protein separation by iso-electrofocusing depends on the charge of the sample ions while on a gel gradient separation depends mainly on the size of the macroions.

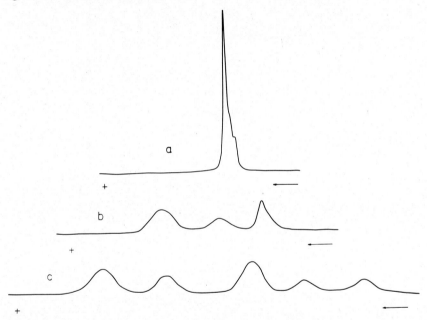

Fig. 5. Separation of LDH isoenzymes from rat brain on 5 μl capillary gels. a: Gel gradient, b: 10% gel, c: Isoelectrofocusing.

Capillaries used: 1, 2, 5, 10 μl Drummond Microcaps*
                  5, 10 μl Brand Micro Pipettes*

Running buffer (B)   0.05 M Tris
                   + 0.384 M Glycin
                   + H₂O ad 500 ml, pH 8.4

* All capillaries (new ones included) should be cleaned by KOCl solution and chromic acid.

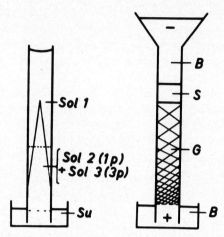

Scheme. The left side shows the migration of solutions 2 and 3 into solution 1 immediately after filling the capillary with gel solutions. The right side shows the schematic arrangement of electrophoresis. B: Running buffer, S: Sample, G: Gel gradient.

Solution 1   35 mg Persulfate + 50 ml water

Solution 2   3.4 g Tris
             + 6 ml 1N H_2SO_4
             + 0.06 ml Temed
             + 6N H_2SO_4 ad pH 8.6
             + H_2O ad 10 ml

Solution 3   20 g acrylamide
             + 0.4 g Bis-acrylamide
             + 3.0 mg $K_3Fe(CN)_6$
             + H_2O ad 37.5 ml

Keep all solutions at 4°C but use them at room temperature. Solution 1 has to be replaced every three weeks. While polymerizing the capillary has to be kept in a wet chamber for about 90 minuts. Afterwards the capillaries with the gel can be stored for about one week in gelbuffer at 4°C.

References

(1)   Kopperschläger, G., W. Diezel, B. Bierwagen und E. Hofmann, FEBS Letters 5, 221–224 (1969).
(2)   Waehneldt, T. V. and P. Mandel, Brain Res. 40, 419–436 (1972).
(3)   Allen, R. C. and H. R. Maurer, Disc Electrophoresis. Walter de Gruyter, Berlin (1971), p. 29.
(4)   Quentin, C. D. and V. Neuhoff, Intern. J. Neuroscience 4, 17–21 (1972).

7.4 Simultaneous Preparation for Electrophoresis of a Large Number of Micro Polyacrylamide Gels with Continuous Concentration Gradients

W. Dames and *H. R. Maurer*

The advantages of using continuous polyacrylamide gradient gels for the electrophoretic separation of very complex protein mixtures have been widely recognized (1–4). A method to produce continuous gradient gels for the usual macroscale was developed by Margolis and Kenrick (1) who adopted and modified a similar well-known technique of biochemistry. The method uses two connected vessels filled with different acrylamide solutions and a mixing device. Following mixing the gradient solution is filled into the gel mold from the bottom to the top. This method, however, cannot be scaled down to the micro scale using capillary tubes since the capillary attraction prevents a slow and smooth filling of the tubes with the mixed gradient solution.

We have overcome this problem by first filling and completely embedding the capillary tubes with buffer solution and then slowly replacing the buffer solution by the developing gradient solution moving from the bottom to the top. Our method allows the simultaneous preparation of a large number of identical micro gradient gels in capillary tubes of 1–50 μl volume.

Materials

Stock solutions:

A: Tris   48.5 g
    Water to   90   ml
    H_2SO_4 98%   3.0   ml
    Temed   0.65 ml
    H_2SO_4 to pH   8.9
    Water to   100.0 ml

B: Stock solution A   1   ml
    Water   7   ml

C: Acrylamide   26.14 g
    Bis (C = 2%)   0.54 g
    $K_3Fe(CN)_6$ 6.0 mg
    Water to   50.0 ml

Electrode buffer:

Tris   6.0 g
Glycine   28.8 g
Water to   1000.0 ml
                                    (pH 8.5)

Sorbitol solution:

Sorbitol   50.0 g
Stock sol. B to   100   ml

Glass capillaries:
  Drummond Microcaps of
  1, 2, 5 μl volume
  Silicone or PVC tubings of
  2 mm i. d.

D: Ammonium =   580.0 mg
   persulfat $(NH_4)_2S_2O_8$
   Triton X-100 4% aq.
                    to 100.0 ml
   Triton may be omitted if the
   highest gel concentration T is
   less than 25%

Polyethylene tubing: < 1 mm i. d.

Cellulose nitrate centrifuge tubes
   (Beckman Instruments):
   Size 9/16", dia 3¾, (No. 33101)

Methods

Assembly of devices

Figure 1 and 2 show the set up for preparing polyacrylamide gradient gels in
micro capillaries. It comprises a mixing device connected with a chamber for

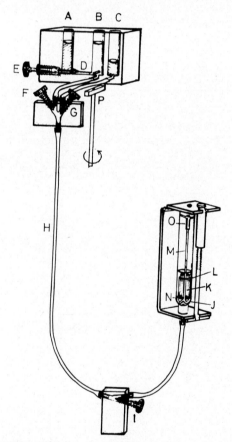

Fig. 1. Set up for the simultaneous preparation of micro polyacrylamide gels with linear-
continuous concentration gradients. For explanation see text.

filling the capillaries. The mixing device (gradient former, figure 2a) consists of 3 chambers, two of which (A and B) are connected at the bottom by a groove (D) which may be closed with a needle valve (E). Chamber A contains the high concentration gel solution, B the low concentration gel solution, and C the sorbitol solution. The gradient former is placed on a magnetic stirrer (P) with a magnetic stirring bar in chamber B. The effluents from chambers B and C are controlled by the two-way stop cock valves F and G. This allows closure of each tubing separately. The common outlet is connected with a one-way stop cock valve and with the tube leading to the filling chamber (gradient tube, figure 2b). This cellulose nitrate tube has a small bottom hole and contains the capillaries (K). The capillaries sit on a nylon mesh disk which has a small central hole and is placed horizontally on the 3 legs of a stainless steel holder (M). A small glass cylinder (L) bundles the capillaries together. A T-piece (O) is put on top of the holder (M) to maintain the capillaries in an upright position. The bottom of the tube is covered with a perforated rubber stopper containing a 2 mm cannula. The top of the tube is closed with a rubber lid.

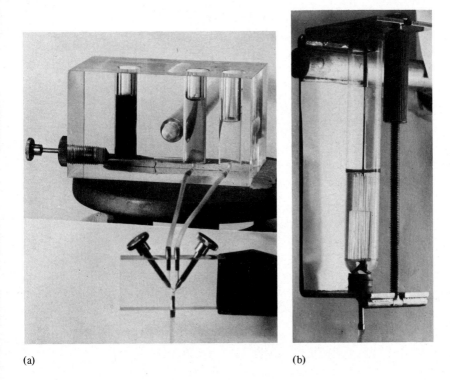

(a)   (b)

Fig. 2. Mixing device with stop cock (a) and filling chamber (b). Compare figure 1. For explanations see text.

To fill the tube system all valves are closed. Some sorbitol solution is placed
into chamber C and the valve G is slowly opened to expell the air from the polyethylene tubing between C and G. The valve G is then closed. Chamber B is
filled with solution B and the valves F and I are opened to allow flow along
tubing H into the filling chamber. Tubing H should have a relatively large i. d.
to compensate the lack of buffer due to the abrupt suction of all capillaries
once they contact the liquid surface. No air should be aspirated by the
capillaries not to avoid the flow into the capillaries. When completely filled
the valves F and I are closed. At this stage the capillaries contain and are
immersed in buffer solution. Thus capillary attraction is no longer an effect.
The gel gradient solution may continuously enter the capillaries from the
both inside and outside the capillaries. Finally tubing H of large i. d. should be
changed for a tubing of small i. d. ($<$ 1 mm) to reduce phase transition of different densities in the descending part of the tubing system to a minimum.

Preparation of the gel solutions

Gel gradients from 2 to 40% may be formed by mixing the stock solutions as
listed in table 1. Two gel solutions of different acrylamide concentrations are
prepared from:

Stock solution C	0.3 to 6.0 vol. parts (= 2–40%)
dist. water	5.7 to 0 vol. parts
stock solution A	1 vol. part
stock solution D	1 vol. part

See legend to table 1 for detailed procedure.

Preparation of the gradient gels

The low acrylamide concentration solution is filled into chamber B. Groove D
is also filled with this solution by quickly withdrawing and closing the valve E.
The magnetic stirrer is started and an equal amount of the high concentration
solution with a trace of bromphenol blue is added to chamber A. Valves
F and I are gently opened. Then needle valve E is opened care being taken
that no diluted gel solution enters chamber A. The mixed gradient solution
should elute at a slow and constant flow rate (about 0.2 ml/min.). When the
chamber A and B are emptied stop cock F is closed and G is opened to allow
the sorbitol solution flushing the tube system until is reaches the lower end of
the capillaries in the filling tube to a level just touching the nylon mesh disk.

Gradient mixing time and filling the capillaries should not exceed 15 min.
Polymerization starts from the top after about 25 min. and has reached the
bottom after about 50 min. Following polymerization the steel holder and

Table 1. Composition of gel solutions for the preparation of defined polyacrylamide concentration gradients. To prepare 2 ml of gel solution, the given (calculated) ml quantity of stock solution C is mixed with the given (calculated) ml quantity of water and 0.25 ml of stock solution A and 0.25 ml of stock solution D are added.

% Acrylamide Concentration	Quantity (ml)		% Acrylamide Concentration	Quantity (ml)	
	Sol. C	Water		Sol. C	Water
1	0.0375	1.4625	21	0.7875	0.7125
2	0.075	1.425	22	0.825	0.675
3	0.1125	1.3875	23	0.8625	0.6375
4	0.15	1.35	24	0.90	0.5625
5	0.1875	1.3125	25	0.9375	0.5625
6	0.225	1.275	26	0.975	0.525
7	0.2625	1.2375	27	1.0125	0.4875
8	0.30	1.20	28	1.05	0.45
9	0.3375	1.1625	29	1.0875	0.4125
10	0.375	1.125	30	1.125	0.375
11	0.4125	1.0875	31	1.1625	0.3375
12	0.45	1.05	32	1.20	0.30
13	0.4875	1.0125	33	1.2375	0.2625
14	0.525	0.975	34	1.275	0.225
15	0.5625	0.9375	35	1.3125	0.1875
16	0.60	0.90	36	1.35	0.15
17	0.637	0.863	37	1.3875	0.1125
18	0.675	0.825	38	1.425	0.075
19	0.7125	0.7875	39	1.4625	0.0375
20	0.75	0.75	40	1.50	–

the capillaries are removed from the filling tube by blowing air through the bottom hole. The outer gel is pealed off and the capillaries are freed from the glass-cylinder.

Storage of the gels

Depending on the Triton concentration the gel gradients tend to detach from the glass wall of the capillaries on storage. Eventually they may slide out of the capillaries. To prevent this the gels are left for about 10–20 min. at open air to produce a small air bubble on both ends of the capillary due to drying. Then the caps are kept air tight in a small glass vial. Stored in this manner, they maintain their separation qualities for a least 2 weeks.

Electrophoresis

The high molarity of the gel buffer ($\sim$ 500 mM) produces a high voltage gradient thus improving resolution considerably. Heat dissipation is rapid due to the small diameter of the gels.

In continuous buffer systems the sample buffer should be diluted 1:10 to 50 to produce a satisfactory concentration (5). The sample should contain 15% glycerol. In discontinuous buffer systems with discontinuous pH gradients (according to Ornstein) the sample may be dissolved in a suitable stacking buffer (containing 30% glycerol) as a substitute for a stacking gel (6). In discontinuous buffer systems at constant pH (according to Allen) the sample is dissolved in low molarity (5–50 mM) sample buffer containing 15% glycerol. The cap gel according to Allen (7) may be replaced by a cap solution being identical to the gel buffer (stock solution B).

For electrophoresis constant voltage between 30 and 60 V at 30–80 μA is used.

Results

To visualize the shape and reproducibility of the gel gradients we have added Blue Dextran 2000 to stock solution C. Following polymerization the gels were removed from the capillaries by water pressure and scanned with a Joyce Loebl double beam microdensitometer. Figure 3 demonstrates that gels of different dimensions (of 1, 5 and 50 μl volume) may be produced in the same filling tube. Figure 4 reveals the excellent linearity of the gel gradients of different acrylamide concentrations produced in 1 and 5 μl capillaries. Figure 5 presents the high resolution of human serum proteins obtained in 4–40% gel gradients of 1 and 5 μl volume. The high reproducibility of the separations should be stressed. Figure 6 exemplifies the good adaption of the micro gradient gels to the electrophoretic separation of molecules of a wide range of size such as RNAs.

Discussion

The exothermic polymerization reaction in the filling chamber may lead to thermic convection currents with deleterious effects on the gradient. To avoid turbulence the gel solution must polymerize from the top to the bottom (1). The addition of potassium ferricyanide slows down the polymerization rate much less in the solution of low acrylamide concentration than in that of high concentration. The catalyst system N, N, N′, N′-tetramethylethylene-diamine (Temed) and ammonium persulfate is kept at the same concentration throughout in order to obtain a more rapid polymerization in the low concentration end and a progressively slower polymerization as gel concentration increases. Sucrose (10–30%) may be added to aid in stabilizing the gradient.

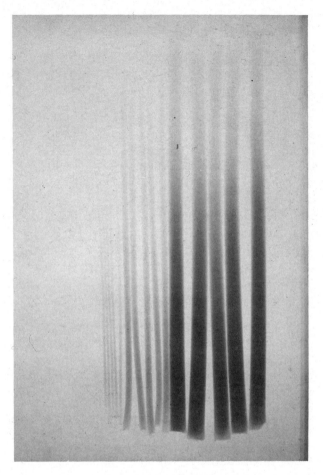

Fig. 3. Micro gradient gels containing Blue Dextran 2000 of 1, 5 and 50 μl volume (left to right) prepared in the same filling chamber. Magn. 3x.

Our procedure to prepare micro polyacrylamide gels with continuous concentration gradients offers several advantageous features which are summarized as follows.

1. Micro gels require less sample quantity, allow the application of higher voltage gradients due to improved heat dissipation, require shorter electrophoresis periods and provide diminished diffusion due to more rapid fixation and staining than gels of usual dimensions. Continuous gradient gels provide considerably improved resolution of sample components. Thus the combination of both methods offers an outstanding separation tool.

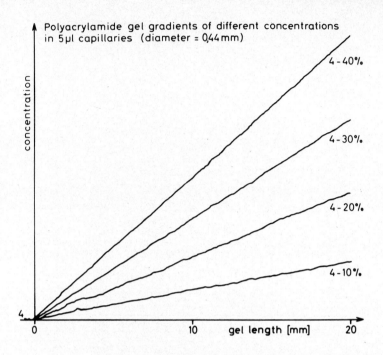

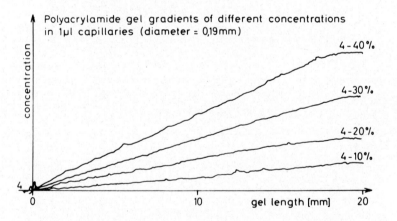

Fig. 4. Densitometer scans of micro gradient gels containing Blue Dextran 2000. Only the first 20 mm of the gels were scanned due to mechanical limits of the densitometer. Grey wedge = 0.39 d, travel service 1:20, magn. 10x.

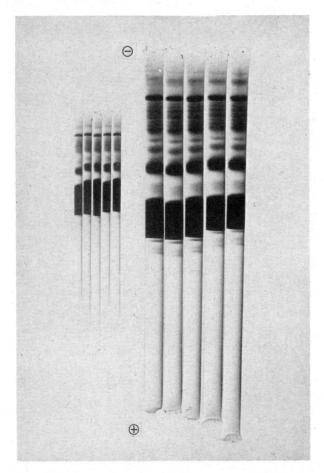

Fig. 5. Human serum proteins separated in 4−40% micro gradient gels of 1 μl (left) and 5 μl (right) volume. Discontinuous buffer system as mentioned in methods according to Allen (7), electrophoresis at 60 V, 70−20 μA, 35 min. Amidoblack staining. Magn. 10 ×.

2. A large number (e. g. as many as 150 5 μl-caps) of gels of either the same or different sizes can be prepared simultaneously. All gels are virtually identical and are automatically layered with buffer to provide a flat surface. Our procedure is easier than the usual technique of micro disc electrophoresis (8) since it avoids several technical steps involving the risk of errors (e. g. water layering, the preparation of stacking gels). We even recommend our procedure when a large number of micro gels of continuous acrylamide concentration are to be prepared due to the excellent reproduceability between gels. If, however, only a few micro gradient gels are needed, another technique may be advised (9).

Fig. 6. R-RNA (0.3 μg of mouse myeloma) separated in a 2–40% micro gradient gel of 5 μl volume containing 5% sucrose. Discontinuous buffer system as in figure 5. Electrophoresis at 30 V, 30–10 μA, 80 min. We thank Dr. Wolfgang Pfeifer for providing the sample of RNA.

3. Different buffer systems may be used as for usual polyacrylamide gel electrophoresis: Continuous and discontinuous buffer systems according to Ornstein-Davis, Jovin or Allen (6).

4. The slope of the gel gradient can be easily varied within the theoretical range between 0 and 40% acrylamide concentration, while in actuality a gradient of 2.0 and 40% T may be achieved.

5. The micro gels can be stored in a cold, humid atmosphere for a least 2 weeks without deterioration of the resulting pherogram resolution. Thus the possibility of simultaneously preparing a large number of well defined polyacrylamide gradient gels on a micro scale specific for various fractionation problems offers a valuable method particularly for complicated separation tasks.

Summary

A method is described to prepare simultaneously a large number (e. g. up to 150 5 μl-gels) of micro polyacrylamide gels (of 1, 5, 10 up to 50 μl volume) with continuous concentration gradients of 2.0–40% acrylamide. The features of the technique are discussed.

References

(1)   Margolis, J., and K. G. Kenrick, Anal. Biochem. *25*, 347 (1968).
(2)   Slater, G., Anal. Biochem. *24*, 215 (1968).
(3)   Caton, J. E., and G. Goldstein, Anal. Biochem. *42*, 14 (1971).
(4)   Maurer, H. R., and F. A. Dati, Anal. Biochem. *46*, 19 (1972).
(5)   Hjertén, S., S. Jerstedt and A. Tiselius, Anal. Biochem. *11*, 219 (1965).
(6)   Maurer, H. R., Disc Electrophoresis and Related Techniques of Polyacrylamide Gel Electrophoresis, 2nd edition, Walter de Gruyter, Berlin–New York 1971, p. 22 f.
(7)   Maurer, H. R., and R. C. Allen, Z. klin. Chem. u. klin. Biochem. *10*, 220 (1972).
(8)   Neuhoff, V., Arzneimittel-Forschg. (Drug. Res.) *18*, 35 (1968).
(9)   Rüchel, R., This volume p. 215.

7.5 Identification of Serum Proteins in a Combined Micro-disc and Micro-antigen-antibody Crossed Electrophoresis

Werner Giebel

The quantitative analysis of individual serumproteins in inner ear fluids presents two problems: 1. the minute volume of these fluids; one inner ear yields only 1–3 μl of endolymph or perilymph, 2. the low protein content of the inner ear fluids; 25 mg% in endolymph and 100 mg% in perilymph. Therefore, the individual serumproteins have to be determined simultaneously. We developed a method combining a micro-disc electrophoresis in capillary tubes with a micro antigen-antibody crossed electrophoresis in horizontal agarose gel slabs (Laurell 1965). An application of this new method to guinea pig inner ear fluids capable of determining a protein concentration as low as 0.3 nanogram of protein per fraction from endo-lymph and perilymph of guinea pigs has been reported (Giebel 1972). A similar method was published by Dames et. al. (1972) but the identification of the single peaks was not demonstrated.

The micro-disc electrophoresis was carried out according to Felgenhauer (1972) in capillary tubes of 5 cm length and 1 mm internal diameter. The acrylamide concentration was 7% in the separation gel and 2% in the stacking gel. For the separation the equipment of Boscamp, Hersel, Germany was used. The micro antigen-antibody crossed electrophoresis was carried out in horizontal agarose gel slabs which were prepared by pouring 1 ml of anti-serum containing 0.8% agarose on microscope slides of the size 26 × 38 mm in plastic frames of an inner size of 27 × 40 mm. After the second electro-phoresis the slabs were washed in 0.15 M NaCl solution for 24 hours. After drying the slabs, they were stained with an 0.25% aqueous solution of Comassie Brilliant Blue R 250. For further methodical details, see Giebel et al. (1973).

The different peaks have been identified by adding a small amount of mono-valent antiserum to the polyvalent antiserum which is mixed in the agarose. The effect of this addition is, that only one peak becomes smaller in this preparation compared to the pattern when only polyvalent antiserum is applied. This diminished peak represents the serumprotein of which the monovalent antibody was added. Unfortunately in commercially available polyvalent antisera the titers of the antibodies against the individual proteins are very different. Therefore the sizes of the peaks are so different that they cannot be visualized under a single set of experimental conditions. Therefore, we mixed certain monovalent antisera to the polyvalent antiserum and called this "optimal" polyvalent antiserum. All antisera were purchased by Behring-Werke, Marburg, Germany.

The resulting protein pattern (figure 1) shows 21 peaks of which 18 could be correlated to distinct serum proteins.

References

(1)   Dames, W., H. R. Maurer, and V. Neuhoff, Hoppe Seyler's Z. Physiol. Chem. *353*, 554 (1972).
(2)   Felgenhauer, K., Biochim. Biophys. Acta *133*, 165 (1967).
(3)   Giebel, W., Arch. klin. exp. Ohr.-, Nas.- u. Kehlk. Heilk. *202*, 417 (1972).
(4)   Giebel, W., and H. Saechtling, Hoppe Seyler's Z. Physiol. Chem. *354*, 673 (1973).
(5)   Laurell, C. B., Anal. Biochem. *10*, 358–361 (1965).

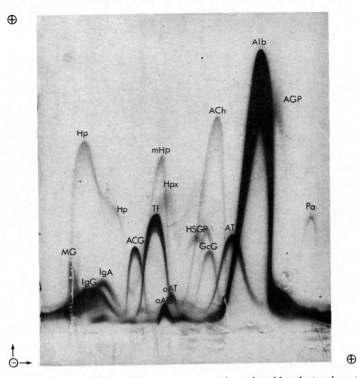

Fig. 1. Protein pattern of 3.3 µg of human serum protein analyzed by electrophoresis in 7 per cent micro-disc gels followed by horizontal micro antigen-antibody crossed electrophoresis in an 0.8 per cent agarose gel slab. Rabbit anti-human antisera to whole sera and serum fractions as follows were added per ml of agarose.

    40 µl antihuman serum
     4 µl antihuman pre-albumin
     4 µl antihuman acid alpha-1-glycoprotein
     8 µl antihuman IgG.

ACG	=	β_1-A/C-globulin, ACH = a_1-antichymotrypsin,
AGP	=	acid-a_1-glycoprotein, Alb = Albumin,
AT	=	a_1-antitrypsin, HSGP = a_2-HS-glycoprotein
GcG	=	Gc-globulin, Hp = haptoglobins, Hpx = hemopexin,
IgA	=	immunoglobulin A, IgG = immunoglobulin G, MG = a_2-macroglobulin,
mHp	=	haptoglobin monomer, oAlb = Albumin oligomere, oAT = a_1-antitrypsin oligomere, Pa = pre-albumin, Tf = transferrin.

Chapter 8. Biochemical Applications

8.1 Electrophoresis in Polyacrylamide-Agarose Composite Gels: An Outline of the Method in Its Applications

José Uriel and Josette Berges

Polyacrylamide gel electrophoresis, independently developped by Raymond (1) and Ornstein and Davis (2), has gained a wide popularity as one of the best methods for the electrophoretic separation of natural substances. Several years ago we described the preparation of acrylamide-agarose composite gels and their application to protein electrophoresis (3, 4). While keeping the excellent resolving power of polyacrylamide gels, the composite gels gain, on account of the addition of agarose, new mechanical properties, elasticity, tensile strenght, ease of handling which enlarge and improve its use particularly at acrylamide monomer concentrations below 5 %. Peacock and Dingman (5) subsequently extended the use of composite gels to the electrophoretic separation of necleic acids, and more recently, the preparation of acrylamide-agarose gels in bed form have enabled us to apply these gels in molecular exclusion chromatography (5 a).

Preparation of composite gels

The principle of preparation of a composite gel is as follows: After having mixed the solutions of acrylamide (monomer and dimer) and agarose, the Poly-merization of acrylamide is made by means of suitable catalysts at a temperature higher than the gelification point of agarose. When then the temperature is lowered, the agarose gelifies in the meshes of the acrylamide gel. By using appropriate concentrations of acrylamide in the mixture, gels of different porosity may be prepared.

Procedure

Whatever the relative composition of the gels, the method of operation for their preparation is always the same.

1. The device represented in figure 1 is placed horizontally in an oven at 55–60° C.

2. The solutions of agarose[1] and that of the acrylamides (monomer and dimer) are prepared in separate vessels taking care that, due to their subsequent mixing in equal parts, the concentration of reagents in these solutions should be the double of that wanted in the gel. The solutions are placed in a water-bath at +50° C.

3. Once the solutions have reached temperature equilibration with the water-bath, the required quantities of both accelerator and catalyst (NNN'N'-tetra-methylene — ethylene — diamine and ammonium persulphate) are added to the agarose solution. The solution is stirred until the ammonium salt is completely dissolved and then the acrylamide solution is added. The mixing should be done rapidly (10 to 15 seconds) but efficaceously and the mixture should be poured down in the device placed in the oven (figure 1).

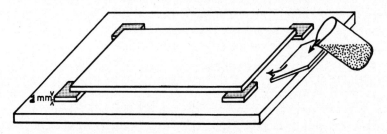

Fig. 1. Device for pouring gels. A standard photographic glass plate (180 × 240 mm) rests on four small plates (2 mm thick) placed upon a flat horizontal surface. The acrylamide-agarose mixture, poured as shown by the figure, fills by capillarity the space between the flat surface and the glass plate.

4. The slab is incubated in the oven for 35 minutes for acrylamide polymeri-zation, then put in a refrigerator or in a cold room (+4° C) for the gelification of agarose (20—40 minutes).

5. The gel is withdrawn from the mould by raising the glass plate and placing it in a dish full of water or of a buffer solution. It is kept cool (+4° C) until use.

Washing the gels before the electrophoretic run is a necessary operation to get rid of the excessive reagents, catalysts and secondary polymerization products of the acrylamide, like acrylic acid. The washing should last at least 16 hours and it is recommended to renew the washing liquid once. It is preferable to

---

[1]   The agarose solution should be made immediately before use by fusion either in a boiling water-bath or on a heating agitator. Also the acrylamide solution should be prepared before use either by dissolution or from an aqueous concentrated stock solution (60% acrylamide, 1.56% bis-acrylamide) kept at 4°−10° C.

wash gels in the buffer solution choosen for electrophoresis. After washing and soaking with buffer solution, the gels are ready for use as electrophoresis carrier. But, they may be kept at $4°$ C for some weeks in the buffer before their use.

Composition of gels

Experience has shown that composite gels in the range of 2.5 to 9 % of acrylamide monomer enables the resolution of the greatest part of non particulate protein molecules.

Table I shows the proportion of various reagents for preparing gels in such range of acrylamide concentrations.
1. The proportions of agarose (0.80 g per 100 ml) and of NNN'N'-tetramethylene-ethylene-diamine (0.06 ml per 10 ml) are the same, whatever the acrylamide content may be.
2. Since ammonium persulphate is a hygroscopic salt, each batch should be divided in small quantities (10 to 20 g) and kept under air-tight cover, preferably at $4°$ C. The figures shown in table I correspond to the use of the same commercial batch. They may differ slightly with other grades.

Acrylamide	Bis-acrylamide	Ammonium persulphate
2,5	0.24	0.04–0.08
3	0.16	0.03–0.06
5	0.13	0.02–0.03
7	0.18	0.01–0.02
9	0.24	0.01–0.02

Concentration: g/100 ml of solution

Acrylamide-agarose gels containing 5–6 % T acrylamide are the most commonly used. In these gels one obtains, in relatively short periods of time (90–120 minutes), good separations of serum proteins, (except a and β-lipoproteins which do not penetrate in the gels).

Gels with 7–9 % T acrylamide are recommended for mixtures of proteins with a molecular weight less than 100.000 daltons. Gels with 2.5–4 % T are applied to the electrophoresis of proteins of high molecular weight (1.000.000 or more).

Gels can be prepared in distilled or demineralized water, or in buffer solutions. Although the proportions of reagents indicated in table I correspond to gels prepared from non-salted aqueous solutions, these proportions are valid when the solutions are replaced by buffer solutions, provided that their pH is between 3 and 8.7 and that their molarity does not exceed 0.3.

The preparation of aqueous non-buffered solutions is the method of choice. These gels can later be soaked in any buffer, without any limitation as to the pH (at least in the range 2 to 10, which we have tested). On the other hand the gel plate may be cut into several strips to be soaked in different buffer. In dehydrated form [2] (see later) these gels may be stored at room temperature and used for electrophoretic runs after rehydration by soaking in the buffer of choice.

Electrophoretic run

One of the advantages of mixed gels is their handiness. Thanks to this, no special apparatus is needed for the electrophoretic separation. The device shown in figure 2 is sufficient. It consists of two rectangular dishes supporting the electrodes and on which lies an horizontal plate supporting the gel plate.

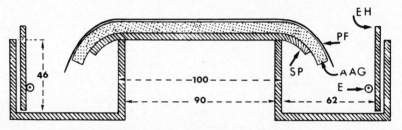

Fig. 2. Device for acrylamide-agarose electrophoresis. A supporting plate SP of glass or plastics is mounted between two dishes C of rectangular cross section. Upon the supporting plate is placed the mixed gel AAG covered with a water-tight sheet PF. The electrodes E of platinum are mounted on removable plates EH. The sizes are in millimeters.

Procedure

Immediately before use, the gel plate is taken out of the buffer solution and quickly dried between two sheets of filter paper. It is then placed immediately on the supporting plate. The length of the gel (figure 1) should exceed by 4–5 cm the length of the supporting plate. We use, as a rule, glass supporting plates of 8 and 10 cm for gels of 12 and 15 cm length, respectively.

The starting wells are cut by means of a punch with sharp edges (two razor blades, 10 to 15 mm long, mounted side by side with a gap of about 2 mm). The minimum distance between two wells should be 16 mm. The wells placed at both gel sides should be distant of not less than 15 mm from the plate edge. The position of the wells with regard to the anodic or cathodic ends of the gel depends on the electric charge of the proteins to be analysed and has to be determined experimentally.

[2]   INDUBIOSE PLATES. Commercially available from Industrie Biologique Française. Gennevilliers – France.

The well is filled with the sample previously hand mixed with molten agarose. With wells of 10 X 2 mm and gels 2 mm thick, it is possible to insert about 0.04 ml of the sample-agarose mixture.

In practice, we operate in the following manner: An aqueous solution of 1.6% molten agarose is kept in a water-bath stabilized at 40–41° C. The sample to be tested is brought either by dissolution or by dilution to a final protein concentration of 1 to 10 μg/ml. An aliquot of the sample is placed in a test tube and immersed into a water bath at 40–41° C. After a few seconds, an equal volume of agarose solution is added to the tube. The mixture is then homogenized by means of a Pasteur pipette with which the starting well is also filled.

The gel is then covered with a thin plastic sheet[3] to avoid water loss due to evaporation during the electrophoresis. Then gel and supporting plate are placed between the two electrode dishes, filled with buffer solution. The junction between the gel and the buffer is made through the ends of the gel strips overhanging the edges of the supporting plate and dipping into the buffer.

A voltage of about 10 volt/cm is applied for the gels soaked with buffer solutions of a conductivity between 1/700 and 1/800 ohm^{-1} cm^{-1}. The duration of electrophoresis for a given sample depends on the acrylamide concentration of the mixed gel.

Buffer selection

As a rule, good electrophoretic separations are obtained if the conductivity of the buffers used is between 1/700 and 1/900 ohm^{-1} cm^{-1}. Conductivities over 1/640 ohm^{-1} cm^{-1} produce by progressive heating some shrinking of the gel, which results in distortion of the electropherogram. A low conductivity is less troublesome but requires an increase in the voltage and/or the duration of the electrophoresis is prolonged.

For general purposes, we use a discontinuous TRIS-glycine buffer, pH 8.7, of the following composition:

    a) Buffer in the gel:

TRIS	9 g
Glycine	4.68 g
H$_2$O	900 ml
1 N HCl to obtain pH 8.7	

    b) Buffer in the electrode vessels:

TRIS	1.2 g
Glycine	25.0 g
H$_2$O	900 ml

---

[3]   "Saran Wrap" sheet – Dow Chemical Co (Midland, Mich. USA).

Demineralized water is added to each buffer to a final volume of 1 litre, the pH is checked and adjusted if necessary. The conductivity of these buffers is about $1/700$ ohm^{-1} cm^{-1}.

Protein staining

Stock solution : 0.25 % Coomassie Brilliant Blue R 250 (6) in 50 % methanol and 10 % acetic acid. The working solution which stores well at room temperature is prepared by dilution of the above (1 : 20) with an aqueous mixture of 40 % methanol, 5 % acetic acid and 2.5 % glycerol.

After electrophoresis, the gels are soacked for at least four hours in the diluted stain solution. The gels are then washed until complete clearing of the background occurs in the 20 % methanol containing 5 % acetic acid and 2.5 % glycerol to which are added several hundred milligrams of an exchanger resin (Dowex type 1 X 2) per liter of washing solution. The resin fixes the stain that diffuses off the gels, thus speeding up the washing procedure.

After washing, the gels, layered on a glass plate, are covered with a cellophane film and can be dried in air. Permanent records and ease of storage are thus obtained.

Acrylamide-Agarose Electrophoresis of Enzymes

General

The flat shape of composite gels and their mechanical properties are particularly appropriate for the use in enzyme electrophoresis. The different steps of manipulation required by the characterization techniques of enzymes after the electrophoretic run are facilitated by the ease of handling of these gels. Moreover their thickness (2—3 mm) which can be reduced by air drying to a thin transparent film favours the photometric scanning of enzyme coloured patterns.

With the aid of appropriate chromogenic substrates many enzymes particularly in the class of hydrolases and oxidoreductases can easily be identified after electrophoresis in acrylamide-agarose. Among them , the following have been studied by this method: carboxylic esterases (7), acetylcholinesterases (8) alkaline phosphatases (9), chymotrypsins (10), lactate dehydrogenases, glucose-6-phosphate dehydrogenases (11), aldolases and peroxidase (12). An example of G 6 PD is illustrated in figure 3.

The characterization reactions for these enzymes will not be detailed here (see references). The procedures are generally simple adaptations of techniques al-

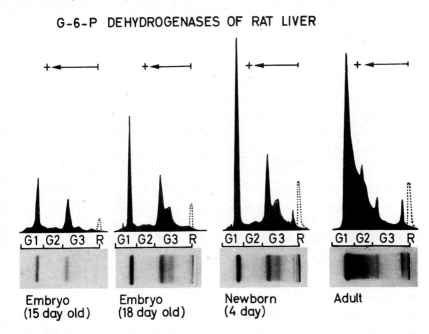

G-6-P DEHYDROGENASES OF RAT LIVER

Fig. 3. Glucose-6-phosphate dehydrogenase transitions in developmental rat liver. Acrylamide monomer, 5%; agarose, 0.8%. Electrophoresis in TRIS-glycin buffer pH 8.7 (from Uriel, reference 11).

ready described for more conventional electrophoresis methods such as starch or agarose.

Two special techniques developed for acrylamide-agarose gels and adaptable to general application are described below.

Characterization of natural inhibitors of enzymes

This technique allows the demonstration and localization after gel electrophoresis of natural inhibitors present in biological solution (biological fluids, tissue extracts, enzyme preparations, etc). The principle is the following: After electrophoresis of the sample presumed to contain inhibitory activity, the gel slab is incubated in a solution of the appropriate enzyme which then enters the gel by diffusion and forms a thin and homogeneous layer on its surface. The slab is then transferred into another solution containing a chromogenic substrate for the enzyme used in the assay. Under the catalytic activity of the enzyme the whole surface of the slab will appear coloured excepting the areas where the inhibitor is present (13).

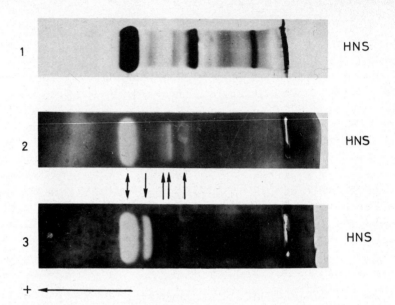

Fig. 4. Characterization of trypsin and chymotrypsin inhibitors in normal human serum.
1. Stained protein bands; 2. inhibitory activity against trypsin; 3. inhibitory activity against
chymotrypsin. (from J. Uriel and J. Berges, reference 13).

Figure 4 illustrates the application of this technique to the characterization of
trypsin and chymotrypsin inhibitors in human serum. Three samples of a fresh
normal human serum were electrophoresed in 5 % acrylamide-0.8 % agarose.
After electrophoresis, one of the slabs (Fig. 4—1) was fixed and stained for
proteins as indicated above, the two others were incubated at 37° C with fresh
solutions containing either trypsin (0.04 mg/ml) or chymotrypsin (0.02 mg/ml),
in 0.1 M phosphate buffer pH 7.4. After 10 min., the plates were transfered to
another tray protected from evaporation by a cover and allowed to stand at
37° C for 30 min. The gel slabs were then immersed in a solution of acetyl-DL-
phenylalanine-β-naphthyl-ester prepared as follows: 5 mg of the substrate in
2 ml of dimethyl-formamide was brought to 20 ml with 0.05 M phosphate buffer
at pH 7.4 in which 10 mg of tetrazotized ortho-dianiside was dissolved. The
mixture was used at once. After 40—60 min. of incubation at room temperature, the gel slabs were removed from the substrate and washed for 30 min in
2 per cent acetic acid.

Along the electrophoretic pathway, the zones possessing inhibitory activity
against the chosen enzyme appear as an unstained band or bands on a coloured
background. The ester is hydrolysed by both trypsin and chymotrypsin, to
yield a coloured product. Where the enzymes are complexed with an inhibitor

the substrate is not hydrolyzed and the resulting bands are not coloured. Figure 4 demonstrates four and two bands with inhibitory activity against trypsin (slab 2) and chymotrypsin (slab 3), respectively.

The same principle can easily be extended to more general applications with many other natural inhibitors, as long as the appropriate chromogenic substrates are available.

Characterization of DNA and RNA polymerases

Catalytic reactions cannot take place within the polyacrylamide gels in the case of enzymes which act on substrates of high molecular weight since large molecules, such as DNA, RNA, or glycogen, are unable to diffuse into the dense network of these gels.

The procedure reported below was developed to overcome this limitation. It consists of the use of two gels with different sieving properties: a small-pore gel (acrylamide-agarose) for electrophoresis and a large-pore gel (agarose) for the catalytic reaction (14).

1. Reagent plate (agarose).

This consits of a thin layer of agarose gel (1 mm thick) on a photographic glass slide. The gel includes the incubation medium (substrates and reagents) for the characterization reaction.
a) *Incubation medium for RNA-polymerase.* Prepare just before use a solution containing: 100 mM potassium maleate (pH 7.5); 8 mM $MnCl_2$; 8 mM β-mercaptoethanol; 08 mM each ATP, GTP, CTP and UTP: 1 μCi ATP-C^{14} ml; and 200 μg DNA/ml.
b) *Incubation medium for DNA polymerase.* Prepare a solution containing: 140 mM Tris-HCl (pH 7.4); 6 mM $MgCl_2$; 2 mM β-mercaptoethanol; 0.06 mM each dATP, dGTP, dCTP, and TTP; 0.4 μCi dCTP-C^{14} ml; and 400 μg DNA/ml.

Heat solution (a) or (b) or (c) to 42° C in a water bath and mix with an equal volume of a 2.4 % aqueous solution of melted agarose kept at the same temperature. The mixture is poured in a device similar to that shown on figure 1, and allowed to gel at room temperature. The reagent plate (glass slide and agarose layer) is then removed from the support plate and covered until use with a plastic film (Saran Wrap) to prevent water loss.

2. Characterization Reaction

The major steps of the experimental procedure are schematically illustrated in figure 5.

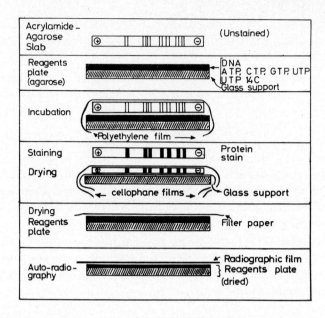

Fig. 5. Schematic illustration (side view) of major steps for characterization of RNA polymerase (see details in text).

Washing. After electrophoresis the gel is immersed for 10 min in the same buffer used to prepare the incubation media (see "Methods"), to equilibrate the gel at the pH optimum for the enzyme reaction.

Incubation. The washed gel is carefully laid on the appropriate reagent plate (figure 9). The "sandwich" is then covered with a plastic film (Saran Wrap) and placed in an incubator at 37° C for approximately 2 hrs. Under these conditions enzymes and other proteins diffuse from the electrophoresced gel into the reagent plate. At the sites to which the polymerases have migrated there is incorporation of labelled as well as unlabelled nucleotides into newly synthesized nucleic acid.

Protein Stain. After incubation the acrylamide-agarose gel slab is removed from the reagent plate and stained for protein with Coomassie blue, as described above.

Autoradiographic Assay. The reagent plate is washed 2 or 3 times for 1–2 hr. in a bath of 2 % aqueous streptomycin. This step ensures the fixation of the polynucleotide newly synthetized, and removes excess of free nucleotides and all other chemicals and reaction products still present in the gel at the end of the incubation period.

The reagent plate is then dried under filter paper and the surface of the dried gel put in contact with a Kodirex film (Kodak) for 1 to 2 weeks at room temperature. The film is developed in the usual way.

References

(1) Raymond, S., and L. Weintraub, Science *130*, 711 (1959).
(2) Ornstein, L., and B. J. Davis, J. Histochem. Cytochem. *7*, 231 (1959).
(3) Uriel, J., and J. Berges, C. R. Acad. Sci. Paris *262*, 164 (1966).
(4) Uriel, J., Bull. Soc. Chim. Biol. *48*, 969 (1966).
(5) Peacock, A. C., and C. W. Dingman, Biochemistry *7*, 668 (1968).
(5 a) Boschetti, E., R. Tixier and J. Uriel, Biochimie, *54*, 439–444 (1972).
(6) Fazekas de St. Groth, R. G. Webster, and A. Datynen, Biochem. Biophys. Acta *71*, 377 (1963).
(7) Failly, Ch., and J. Uriel, Biochim. *54*, 409 (1972).
(8) Unpublished results
(9) Bettane, M., and M. Stanislawski, Biochim. Paris *54,* 209 (1971).
(10) Pacaud, M., and J. Uriel, Eur. J. Biochem.*23*, 435 (1971).
(11) Uriel, J., Path. Biol. vol 17, 877–884 (1969).
(12) Avrameas, S. and T. Ternynck, Immunochemistry *8*, 1175 (1971).
(13) Uriel, J., and J. Berges, Nature *218,* 578 (1968).
(14) Uriel, J., and C. Lavialle, C. R. Acad. Sci. Paris *271*, 1802–1804 (1970).
(15) Berges, J., et J. Uriel, Biochimie *53,* 303 (1971).

8.2 Polyacrylamide Gel Electrophoresis of RNA

E. G. Richards and *R. Lecanidou*

Introduction

By choosing a gel of appropriate composition, RNA molecules in the molecular weight range of any naturally occuring species will migrate in polyacrylamide gel electrophoresis, and by using gradient gels mixtures of species ranging from tRNA to high molecular weight ribosomal RNA may be analysed on the same gel.

Although polyacrylamide gel electrophoresis can be used preparatively for RNA perhaps its most important use is in analytical applications. Such applications may involve the detection of degradation products, molecules with altered conformation or impurities; the quantitative or semiquantitative assay of the proportions of components in mixtures; and the estimation of molecular weights or related parameters.

Zone width

The resolution of two zones in polyacrylamide gel electrophoresis depends on
the difference between the mobility of the two components and also on the
width of the zones. Thus the narrower the zones the better the resolution.

The factor that ultimately limits the narrowness of the zones is diffusion; this
depends on the square root of the time; it follows that diffusion is most important in the early stage of the experiment, and also that there is little point in
sharpening zones beyond a certain point. In fact if the mobility ratio (ratio of
mobility in the gel to that in free solution) is 0.1 or less — as will generally be
the case if the molecular sieve properties of the gel are to be fully utilised — a
sufficient degree of sharpening is produced as the zone enters the gel. Thus with
5 mm diameter gels, a load of 20 μl of RNA solution (in 5 % sucrose) gives a
column 1 mm high of solution on top of the gel. If the mobility ratio is 0.1
the zone width is 0.1 mm after entering the gel but after 2 hours it has diffused
to a thickness of about 1 mm. Even if as much as 100 μl of sample were to be
applied, the zone after diffusion would still not be appreciably greater.

These considerations assume that diffusion is the only zone broadening process
operating. This in general is not so since the front and rear of the zone suffer
from the same sort of effect that lead to boundary anomalies in free boundary
electrophoresis, and which originate in conductivity changes in the vicinity of the
zone. The net result is that the front of the zone is expected to sharpen and
the rear to trail giving assymetric zones as illustrated in figure 1. Not only that,
but the width of the zone increases more rapidly with time than for pure diffusion at a rate depending on the mass of RNA in the zone. This effect is illustrated
in figures 2 and 3. A further effect is that the front of the zone moves more
quickly than the rear as illustrated in figure 4 and may indeed overtake and
obscure other components present in small amounts and travelling just ahead

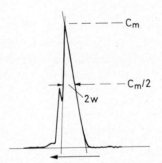

Fig. 1. Profile of a zone containing 5 μg of 5S RNA stained with pyronine Y. The major
peak is renatured 5S RNA and the minor leading peak is denatured 5S RNA.

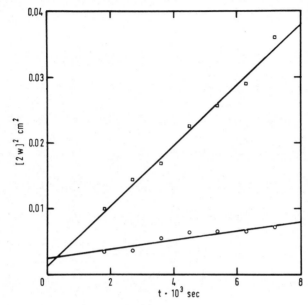

Fig. 2. Effect of load on rate of zone broadening. The load in the upper line was 6.1 μg of renatured 5S RNA and in the lower, 0.9 μg.

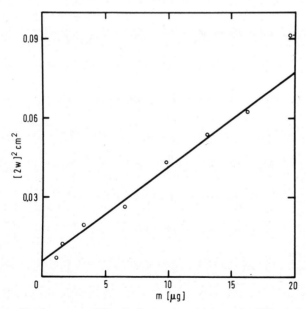

Fig. 3. Effect of load on zone widths. Each experiment was run for 120 min in 0.5 cm diameter gels at 0.028 A/cm^2.

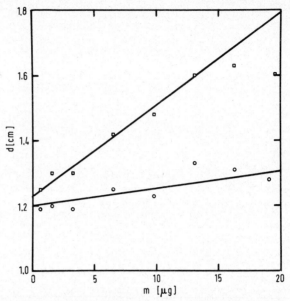

Fig. 4. Plots of distance moved by leading edge (upper curve) and trailing edge (lower curve) versus load of renatured 5S RNA. Each experiment was carried out in the standard system at 0.028 A/cm² for 2 hours.

of a major zone. These effects can be mitigated or eliminated by using small masses of RNA in the load or increasing the buffer conductivity. Thus using 0.05 M tris-HCl as the buffer system the mass of RNA in any zone should not be greater than 1 μg and the theory (1) suggests that for preparative purposes the buffer or electrolyte concentration should be increased in proportion to the load.

Quantitation

Various methods have been proposed for detecting the position of zones of RNA in gels after electrophoresis. It is in fact possible to prestain the RNA with some dye such as acridine orange so that coloured zones may be observed while the run is in progress. Other methods involve the use of radioactively labelled RNA but these are difficult to quantitate. Loening (2) has employed Ultraviolet densitometry of the gels — a technique which is sensitive — 0.05 μg should be easily detectable — but purification of the gel components to eliminate absorbing impurities is involved and the method is rarely applicable to the higher concentration gels required for use with low molecular weight RNA.

We use a staining procedure whereby the gel is soaked in a dilute solution of Pyronine Y in 1 % acetic acid. The dye appears to form an insoluble and stable

complex with the RNA and prevents the zones from diffusing. We have investigated 20 or so similar dyes and have selected pyronine Y as the most effective,
though others such as Azure A and Toluidine Blue are nearly as good. The excess dye is removed electrophoretically as described by Schwabe (3) but we have
found it virtually impossible to remove the dye trapped by the RNA in the
zone. Figure 5 shows absorbance spectra of free Pyronine Y in a gel and of the dye

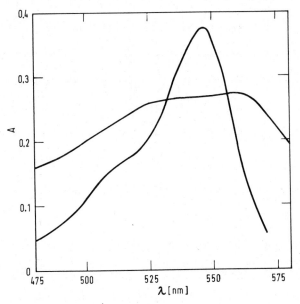

Fig. 5. Spectra of pyronine Y in a 10 % gel (narrow curve) and pyronine Y bound to tRNA in
a 10 % gel (broad curve).

bound to 5S RNA in a similar gel. It may be seen that the absorption band of
bound dye is considerably broadened and this is probably due to the formation of
stacked arrays of dye molecules bound to the RNA. Unfortunately we do not
know either the binding ratio of dye to RNA (and this may vary from RNA
species to RNA species) or the absorbtivity of the dye in the bound form (which
may not be constant either), so absolute estimations of the concentrations of
RNA in gels by this means is at present not possible. Nevertheless the absorbance
is proportional to the RNA concentration for a given RNA species under
standard conditions. The absorbance profiles of stained zones may be determined
in the Gilford gel scanning attachment or alternatively by a Joyce-Loebl microdensitometer; with this latter instrument it is important to use a complementary
filter in the two light beams — and its efficacy depends on the fact that the absorbance of the bound dye is nearly constant over the waveband transmitted

by this filter. We have used this technique for estimating the proportions of
renatured and denatured 5S RNA in a mixture and figure 6 shows a plot of the
proportion estimated in this way versus the proportion in made up mixtures
analysed on gels. The method is of great sensitivity in that quantities of RNA as
little as 0.01 μg are easily detectable and 0.1 μg amounts may be estimated with
some precision.

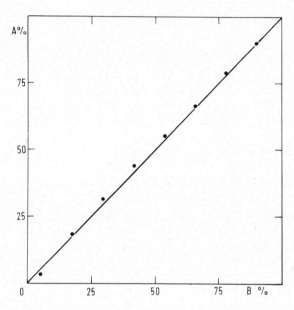

Fig. 6. Proportion of denatured 5S RNA in mixtures of denatured and renatured material
(A) versus proportion estimated from area in scan of stained zones (B).

Molecular weight-mobility relations

The mobility of RNA species in polyacrylamide gels of a given concentration
(T), varies with their molecular weight and various authors have suggested exploiting the fact to estimate molecular weights — usually by measuring the ratio
of the migration distance of a series of RNA species to that of a marker substance
(often bromophenol blue) and plotting this ratio against the logarithm of the
molecular weight to produce a calibration curve. (See (1) for references).
Alternative procedures involve plotting the sedimentation coefficient against
the mobility and utilizing the sort of emprical relation between molecular
weight and sedimentation coefficient proposed by Boedtker et al. (4).
These procedures are at best semi-quantitative since the mobility of an RNA

species depends on other factors besides its molecular weight. This is examplified by the work of Fisher and Dingman (5) and in figure 7 which shows a mixture of denatured and renatured 5S RNA separated on a 16 % gel; no fewer than 3 zones can be discerned; all these conformers have the same molecular weight but differ in their secondary structure.

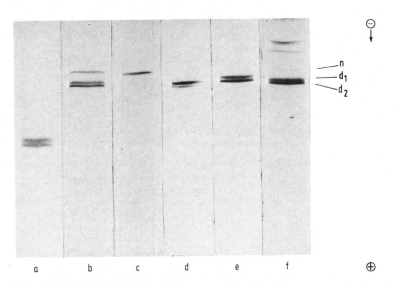

Fig. 7. Photographs of gels containing various conformers of 5S RNA. Gel (a) was 10 % and the rest 16 %. Gels b, d, e and f clearly show that the denatured material may be resolved into two components.

Staynov, Pinder and Gratzer (6) in our laboratory have shown in preliminary experiments that this difficulty may be overcome by running the gels in pure formamide; in this medium — which has a high dielectric constant and is an ionising solvent — secondary structure either due to base pairing or stacking is essentially eliminated and apparently a unique relation holds between mobility and molecular weight for RNA and fractionated homo-polynucleotides. The final proof of this supposition demands accurate knowledge of the true molecular weights of a variety of RNA species, a problem of some difficulty which is under investigation at present in our laboratory.

Finally it has been shown (1) that the mobility depends on the gel concentration (T) and the proportion of cross linking agent (C). In the latter case the mobility measured at constant T, passes through a minimum at about C = 5 % and in the former there is a linear relation between the logarithm of the mobility measured at constant C and the gel concentration as illustrated in figure 8. This relation

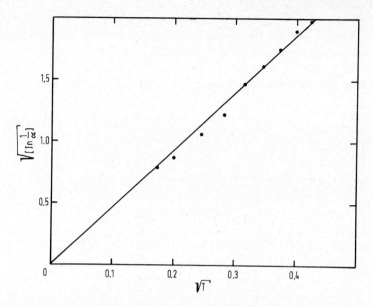

Fig. 8. Plot showing relation between mobility ratio (ratio of mobility in gel to that in free solution), a, and gel concentration T.

is similar to that found by Morris and Morris for proteins (7). The practical consequence of this is that the resolving power increases as the gel concentration increases and for this reason we used 16 % gels to resolve denatured and renatured 5S RNA; these conformers are barely resolved in 10 % gels.

References

(1)   Richards, E. G. and R. Lecanidou, Anal. Biochem. *40*, 43 (1971).
(2)   Loening, U. E. Biochem. J. *102*, 251 (1967).
(3)   Schwabe, C. Anal. Biochem. *17*, 201 (1966).
(4)   Boedtker, H. J. Mol. Biol. *2*, 171 (1960).
(5)   Fisher, M. P. and C. W. Dingman, Biochemistry *10*, 1895 (1971).
(6)   Staynov, D. Z. J. D. Pinder and W. B. Gratzer, Nature New Biology *235*, 108 (1972).
(7)   Morris, C. J. O. R. and P. Morris, Biochem. J. *124*, 517 (1971).

8.3 Molecular Weight Determination of t-RNAs and Their Fragments by Electrophoresis in Non-aqueous Gels

D. Staynov, B. Beltchev, M. Yaneva and W. B. Gratzer

We have presented in a previous paper (1) a method for the determination of the molecular weight of RNA by electrophoresis in non-aqueous acrylamide gel systems. Electrophoretic mobility in such gels appears to be a function of molecular weight (MW) only, and not of the nucleotide composition of the RNA. (A problem to be answered from this work is what is the nature of this function.) It has been shown previously that in the region of 0.5 to 1.5×10^6 Daltons a linear relationship between log MW and electrophoretic mobility yields a good molecular weight approximation (for ribosomal RNAs), and that this function is not linear for molecular weights greater than 2×10^6 Daltons, for example, Tobacco mosaic virus RNA.

In this report, the relationship between MW and electrophoretic mobility in the region of 3 to 40,000 Daltons, the region of tRNA's and their fragments, is reported. In this molecular weight range, the polynucleotide chain length is short and consists only of a few statistical segments. This may lead to some additional factors, affecting the relationship between MW and electrophoretic mobility.

The tRNA's and the fragments, used in this investigation, were obtained as follows:

1. tRNA^{Phe} from yeast was purified by the method of Wimmer et al. (2)
2. tRNA^{Met} and tRNA$_F^{Met}$ (E. coli) were purified as described by Seno et al. (3)
3. $3'-$ and $5'$-halves of tRNA^{Phe} (yeast) were obtained by chemical splitting of the molecule in the anticodon loop sa described by Philippsen et al. (4), separated by DEAE-cellulose column chromatography (5).
4. Fragment $1-18$ of tRNA^{Phe} (yeast) from $5'$-end, a gift from Dr. Thiebe and was obtained as described in ref. 6.
5. Fragment $1-15$ of tRNA^{Phe} (yeast) was obtained and separated as previously described (7).
6. Oligo G ($10-12$ nucleotides) were obtained as described in ref. 8 and characterized by end group analysis.

Polymerization of acrylamide, purification of formamide, and electrophoresis were performed as described in ref. 1. Staining was not carried out with Pyronin Y because of its ineffectiveness in precipitating small fragments of RNA. For this purpose we used 1-ethyl-2(3-(1-ethylnaphtho(1,2 d) thiazolin-2-ylidene)-2-methylpropenyl)-naphtho(1,2 d) thiazolium bromide (Stains All) from Eastman Organic Chemicals Distillation Products Industries, Rochester 3, New York, Division of Eastman Kodak Company, No. 2718.

Gels with different concentrations of acrylamide were used — from 7.5 to 15 per cent T (w/v) with 5 per cent C. In all gel concentrations, we observed monotonic functions of MW versus electrophoretic mobility with different curvatures. The shape of the curves log MW — electrophoretic mobility, depending upon the gel concentration, was convex, concave, or S-shaped, but never straight-lined.

In figure 1 is presented an S-shaped curve of log MW versus electrophoretic mobility in 12 per cent T acrylamide gel. The electrophoretic mobilities are calculated relative to that of tRNA^{Phe}.

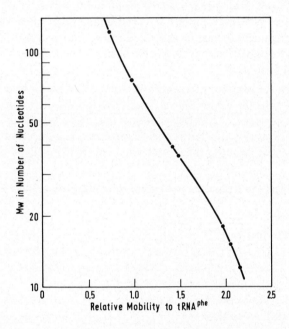

Fig. 1. Molecular weight-electrophoretic mobility relation for tRNA's and fragments in 12 % polyacrylamide gel containing formamide. Each point on the curve corresponds to one of the fractions previously mentioned. Fractions tRNA^{Phe}, tRNA^{Met}, tRNA$_F$^{Met}, and tRNA^{Phe} treated with HCl have the same mobilities and are presented with one point.

All these curves were smooth and no fragment gave any deviation. However, with this system it is impossible to calculate the MW of unknown fragments for two reasons: 1. the lack of a universal curve; 2. poor reproducibility of the gels with acrylamide and formamide from different origins. The only way to determine the MW of such fragments is to obtain curves for known fragments in the same gel. The accuracy in determination of the MW for fragments, smaller than 30 nucleotides, was ± 1 nucleotide, whereas for intact tRNA molecules, this value was ± 2−3 nucleotides.

References

(1)   Staynov, D., J. Pinder, and W. B. Gratzer, Nature New Biology 235, 108 (1972).
(2)   Wimmer, E., I. H. Maxwell, and G. M. Tener, Biochemistry 7, 2623 (1968).
(3)   Seno, T., M. Kobayashi, and S. Nishimura, Biochim. Biophys. Acta 169, 80 (1968).
(4)   Philippsen, P., R. Thiebe, W. Wintermeyer, and H. G. Zachau, Biochem. Biophys. Res. commun. 33, 922 (1968).
(5)   Beltchev, B. and M. N. Thang, PEBS Letters 11, 55 (1970).
(6)   Harbers, K., R. Thiebe and H. G. Zachau, Eur. J. Biochem. 26, 132 (1972).
(7)   Beltchev, B. and M. Grunberg-Manago, FEBS Letters 12, 24 (1970).
(8)   Beltchev, B. and M. Grunberg-Manago, FEBS Letters 12, 27 (1970).

8.4 Conformational Changes of Low Molecular Weight RNA as Visualized by Polyacrylamide Gel Electrophoresis

Georg R. Philipps

In 1964, Rosset et al. (1) described a low molecular weight RNA in *Escherichia Coli* which later was found to be part of the larger ribosomal subunit. This so-called 5 S RNA was subsequently also detected in eucaryotic cells (2,3). It has a chain length of 120 nucleotides (4) and thus is slightly larger than the different tRNA species (75 to 87 nucleotides). In the following years, several other low molecular weight RNA have been described. A 6 S RNA was isolated from the cytoplasm of *E. Coli* (5), it contains 185 nucleotides (6). This RNA has never been found in eucaryotic cells. On the other hand, a 7 S RNA (also called 5.8 S RNA) has been detected in ribosomes of eucaryotic cells where it appears to be hydrogen-bonded to 28 S rRNA (7,8). A number of low molecular weight nuclear RNA's are also known, one of which, a 4.5 S RNA, has recently been sequenced by Ro-Choi et al (9). Some of these nRNA's may well arise from high molecular weight RNA during maturation processes in nuclei but others probably represent genuine nRNA's. Finally, one particular precursor to tRNA[Tyr] has so far been characterized by its sequence (10) and was found to contain 48 nucleotides more than the mature species. However, it can be expected that other pre-tRNA molecules with slightly different chain lengths are present in the cells.

Due to this great variety of low molecular weight RNA it became increasingly more
difficult to purify particular RNA species according to their chain length by gel
filtration on Sephadex. The different low molecular weight RNA's are eluted
from these columns in the fractions which immediately precede tRNA and the
latter is eluted in a rather broad peak which is somewhat dependent on the
buffer used and the distribution of individual species (11). Therefore, a clear
separation of these different RNA's is almost impossible as seen in figure 1.

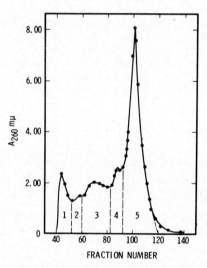

Fig. 1. *Gel Filtration Analysis of Low Molecular Weight RNA.* The RNA was obtained
from *E. Coli* by phenol extraction of the post-ribosomal fraction. It was precipitated with
ethanol and about 50 mg of RNA were applied to a column of Sephadex G. 100 (150 × 2.5 cm)
The RNA was eluted with 10 mM Tris-HCl, 1.0 M NaCl pH 7.2 at a flow rate of 25 ml/h.
Fractions were pooled as indicated by dashed lines.

Furthermore, the distribution coefficient of these RNA's is dependent on the
buffers used for chromatography. Figure 2 depicts the profiles of RNA from
fractions 3 and 4 of Figure 1 upon rechromatography under different conditions.
It can be seen that there are major differences in the elution profiles of the three
minor peaks. Thus it follows that it is impossible to characterize low molecular
weight RNA by gel filtration on Sephadex. Furthermore, since no easy biological
assay is known to specify any low molecular weight RNA with the exception of
tRNA, I searched for other means by which some of these RNA's might be characte
ized. A rather high resolution of different RNA's can be obtained by electrophoresis in polyarcrylamide gels (12–14). The RNA of figure 2 which was
eluted prior to tRNA was thus subjected to this method and the results are

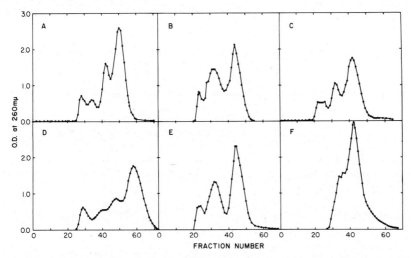

Fig. 2. *Rechromatography of RNA Eluted Prior to tRNA*. To a column of Sephadex G 100 (150 × 1.5 cm) were applied 3 mg of RNA which had previously been separated from tRNA (fractions 3 and 4 of Fig. 1). The columns were equilibrated and run in the following buffers: (A) 10 mM Tris-HCl, 10 mM Mg^{2+}, pH 8.0; (B) 10 mM Tris-HCl, 10 mM Mg^{2+}, pH 4.5; (C) 10 mM acetate, 5 mM EDTA, pH 4.5; (D) 10 mM Tris-HCl, 1 M NaCl, pH 7.0; (E) 5 mM acetate, 5 mM Mg^{2+}, pH 4.8; (F) 5 mM acetate, pH 4.8. All buffers contained also 1 mM $Na_2S_2O_3$. Fractions of 1.5 ml were collected.

shown in figure 3. There are in all samples three major RNA-staining bands: two between 7 and 8 cm and one at about 6 cm. A number of slower moving bands which vary quantitatively and qualitatively in the different samples are also visible. Since I expected to find only one (for 5 S RNA) or two (for 5 S and 6 S RNA) bands, it was surprising to find a third band in all these samples.

In order to determine which of the two bands between 7 and 8 cm represents 5 S RNA, the latter was prepared from ribosomes as described in ref. 15. The last step of this purification was a preparative polyacrylamide gel electrophoresis in TBEM buffer (90 mM Tris/90 mM boric acid/0.2 mM EDTA/1mM magnesium acetate, pH 8.3). Prior to the electrophoresis the RNA had been heated in a buffer which contained 10 mM EDTA at neutral pH. This was done to dissociate possible aggregates. The electropherogram showed two RNA-staining bands which migrated with slightly different mobilities and the RNA of each of these bands was extracted seperately as described in ref. 14. Of both of these RNA fractions one aliquot was heated at neutral pH in a buffer containing either 10 mM EDTA or 10 mM Mg^{2+}. Heating was for 5 min at 60° and the samples were allowed to cool on the laboratory bench. The different samples were then applied to the same electrophoresis slab prepared with TBEM buffer. As can be seen in figure 4, 5 S RNA showed a faster mobility in polyacrylamide gels after

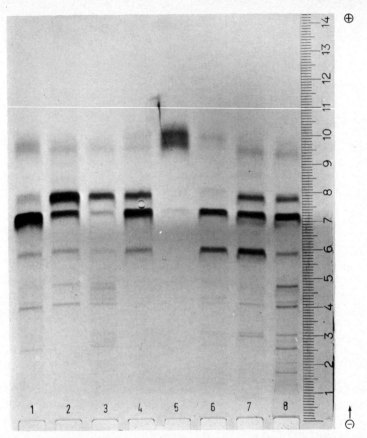

Fig. 3. *Electropherogram of RNA after Rechromatography*. RNA eluted prior to tRNA upon rechromatography (see Fig. 2) was applied to a 10 % polyacrylamide gel using a buffer containing 0.09 M Tris, 0.0032 M EDTA, 0.01 M magnesium acetate, 0.09 M boric acid (pH 8.3). Samples of about 1 A_{260} unit were used and the electrophoresis performed as described in ref. 14. From left to right: (1) RNA from Fig. 2A; (2) RNA from Fig. 2C; (3) RNA from Fig. 2F previously heated in 10 mM EDTA; (4) RNA from Fig. 2D; (6) RNA from Fig. 2B previously heated in 10 mM Mg^{2+}; (7) RNA from Fig. 2F; (8) RNA from Fig. 2C previously heated in 10 mM Mg^{2+}. Migration from bottom to top.

being heated in 10 mM EDTA as compared to 5 S RNA which had been heated in 10 mM Mg^{2+}. However, the same mobility was observed in samples which had been extracted from the faster or slower migrating band of the first (preparative) electrophoresis. This can be seen by comparison of the RNA in slots 1 and 2 of figure 4 (which derived from the faster moving band) with those in slots 3 and 4 (which were from the slower moving band). This showed that there was partially a reversion of the mobility in gels which was only dependent upon the treatment of the RNA prior to the (second) electrophoresis.

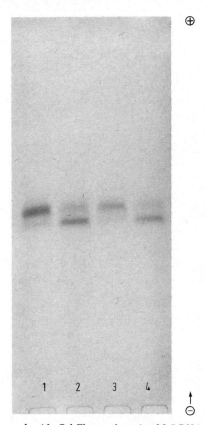

Fig. 4. *Analytical Polyacrylamide Gel Electrophoresis of 5 S RNA*. The RNA was obtained
by purification on a preparative gel electrophoresis after heating in 10 mM EDTA. The
RNA of each of the two resulting bands was extracted seperately and submitted to an
analytical gel electrophoresis. Samples (1) and (2) were from the faster migrating band of
the preparative gel electrophoresis and samples (3) and (4) from the slower migrating band.
Samples (1) and (3) were heated in 10 mM EDTA and samples (2) and (4) in 10 mM Mg^{2+}
prior to the analytical gel electrophoresis.

A similar phenomenon, in this case a change of the elution pattern had previously
been described by Aubert et al. (16). These authors had used urea treatment which
resulted in a change of the distribution coefficient of 5 S RNA after chromatography on Sephadex or MAK columns. They also showed a reversibility of
these changes when the denatured RNA was treated either with Mg^{2+} or 1 M
NaCl. However, the results obtained by Aubert et al and those presented here
probably represent different changes of the molecular structure of 5 S RNA. It
is known that exposure of RNA to high concentrations of urea breaks the secondary and tertiary structure of the latter (17). To study the migration of 5 S RNA

in gels after treatment with urea, 5 S RNA was heated in 7 or 4 M urea
for 5 min at 60° or 90° and electrophoresis was then performed in a gel which also
contained urea. As can be seen in figure 5, again two bands of 5 S RNA can be

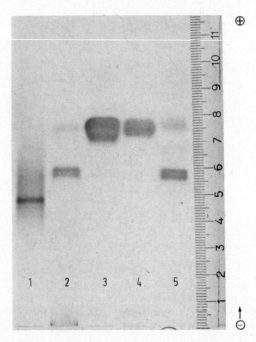

Fig. 5. *Polyacrylamide Gel Electrophoresis of 5 S RNA and 6 S RNA in Urea.* The RNA
used was heated prior to the electrophoresis in 7 M urea for 5 min at 60° and then cooled
quickly. A 7.5 % polyacrylamide gel was used with TBEM buffer containing 4 M urea.
About 1 A_{260} unit of RNA was applied. (1) 6 S RNA purified by gel electrophoresis as
described in ref. 15; (2) 5 S RNA, faster migrating band obtained upon purification by gel
electrophoresis; (3) tRNA from fraction 4 in Figure 1; (4) tRNA from fraction 5 in Figure 1;
(5) 5 S RNA, slower migrating band.

distinguished, however, in this and similar gels the two RNA migrated much
closer together than the two RNA forms in figure 4. I suppose that the two RNA
seen in gels containing urea and obtained after urea-treatment of RNA represent
molecules of slightly different molecular sizes and not two conformational stages.
This conclusion is supported by the observation that the 5 S RNA extracted
separately from the two bands appearing in urea gels and renatured by heating
in Tris-buffer at neutral pH in either 10 mM EDTA or 10 mM Mg^{2+} gave rise
to two bands in both case (figure 6). Thus, the migration of both RNA obtained
in gels with urea depended only on the treatment of the sample prior to electrophoresis in non-urea gels. Earlier, I have proposed (15) that the two bands seen

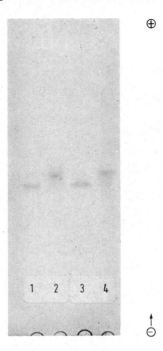

Fig. 6. *Analytical Polyacrylamide Gel Electrophoresis of 5 S RNA Obtained from Gels Containing Urea.* The gel was made as/in Fig. 3. Samples (1) and (2) were from the faster migrating band of the preparative gel electrophoresis in urea and samples (3) and (4) from the slower migrating band. Samples (1) and (3) were heated in 10 mM Mg^{2+} and samples (2) and (4) in 10 mM EDTA prior to the analytical gel electrophoresis.

in urea gels represent 5 S RNA and the precursor to 5 S RNA which have both the same conformation and can undergo the same conformational changes.

Fresco and coworkers (18) and Gartland & Sueoka (19) had observed that certain species of tRNA showed a change of the elution pattern upon Sephadex chromatography when exposed to chelating agents. I have examined six purified tRNA species under the same conditions as described here for 5 S RNA (14). Of these tRNA, only tRNA^{Glu} showed a change in mobility which was comparable to that of 5 S RNA (figure 7). However, since only highly purified tRNA species can be used for these studies it is possible that other species too show a similar behaviour.

No change in mobility was observed when tRNA was aminoacylated. This may reflect either the absence of any conformational change of tRNA after amino-acylation, or a conformational change which cannot be detected by gel electro-phoresis and thus is of a different kind than that described previously.

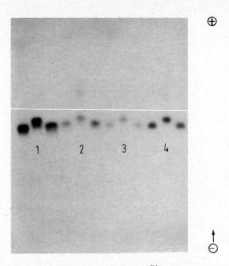

Fig. 7. *Polyacrylamide Gel Electrophoresis of tRNA*^{Glu}. Transfer RNA specific for Glu
was separated into a slower and faster migrating band by preparative gel electrophoresis.
The bands were seperately extracted and analysed in pairs of three. In each pair, the sample
in the first slot remained untreated, the sample in the second slot was heated in 10 mM EDTA
and the third sample was heated in 10 mM Mg^{2+}. From left to right, pair (1) tRNA^{Glu} not
purified by gel electrophoresis; pair (2) tRNA^{Glu} purified by gel electrophoresis but not
separated into slower and faster migrating band; pair (3) tRNA^{Glu}, slower migrating band;
pair (4) tRNA^{Glu}, faster migrating band. (From ref. 14).

Also, 6 S RNA preserves the same mobility when heated in the presence of
EDTA or Mg^{2+}. After treatment with urea 6 S RNA showed only one band in
gels (see figure 5).

On the other hand, a number of RNA which migrated slower than 6 S RNA in
polyacrylamide gels do not represent genuine RNA species but evidently consist
of RNA aggregates. This can be seen in figure 8 where an RNA fraction had
been investigated which was eluted prior to 5 S RNA (see legend to figure 8).
When this fraction was investigated under conditions similar to those described
above it could be seen that upon EDTA treatment most of the slower migrating
bands disappeared. These aggregates do not consist of tRNA since there was only
a slight increase in the intensity of the tRNA band. Instead, the RNA migrates
after EDTA treatment in a broad band in the position of 5 S RNA and between
5 S RNA and tRNA. The possibility exists and preliminary experiments thus
far conducted might suggest that the RNA which apparently can aggregate
with great ease represents precursor molecules of tRNA.

In conclusion, I have shown that 5 S RNA can undergo a conformational change
which can be made visible in polyacrylamide gels. This change is reversible. A

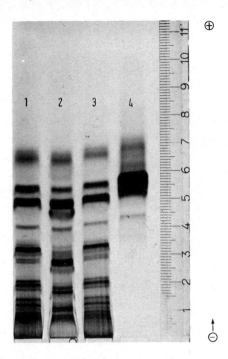

Fig. 8. *Analytical Polyacrylamide Gel Electrophoresis of RNA Aggregates.* The RNA used
in this assay was obtained from an column similar to that in Fig. 1, it was further purified
on BD-cellulose. The Gel was 10 % made in the buffer described in figure 3. Sample (1)
remained untreated; sample (2) was heated for 5 min in 10 mM Tris-HCl (pH 7.0), 10 mM
Mg^{2+} and cooled slowly; sample (3) was heated in the same buffer containing 5 mM glutathione;
sample (4) was heated 5 min in 10 mM Tris-HCl (pH 7.0), 10 mM EDTA.

similar change was observed for $tRNA^{Glu}$. Since there is a lack of an easy assay
for 5 S RNA this change in mobility can be used for the characterization of the
latter.

Acknowledgements

This research was conducted in the Department of Biochemistry, St. Louis
University, School of Medicine, St. Louis, Mo. USA. It was partially supported
by grants from the National Institutes of Health (GM 13364) and the American
Cancer Society (P 511). The assistance of Miss Elisabeth Hugi and Miss Janice
Timko is acknowledged.

References

(1)   Rosset, R., R. Monier and J. Julien, Bull. Soc. Chim. Biol. *46*, 87 (1964).
(2)   Comb, D. G., N. Sarbar, J. DeVallet and C. J. Pincino, J. Mol. Biol. *12*, 509 (1965).
(3)   Galibert, F., C. J. Larsen, J. C. Lelong and M. Boiron, Nature *207*, 1039, (1965).
(4)   Sanger, F., G. G. Brownlee and B. G. Barrell, J. Mol. Biol. *13*, 373 (1965).
(5)   Goldstein, J., and K. Harewood, J. Mol. Biol. *39*, 383 (1969).
(6)   Brownlee, G. G., Nature New Biology *229*, 1947 (1971).
(7)   Pene, J. J., E. Knight and J. E. Darnell, J. Mol. Biol. *33*, 609 (1968).
(8)   Sy, J., and K. S. McCarthy, Biochim. biophys. Acta *199*, 86 (1970).
(9)   Ro-Choi, T. S., R. Reddy, D. Henning, T. Takano, C. W. Taylor and H. Busch,
       J. Biol. Chem. *247*, 3205 (1972).
(10)   Altman, S., and J. D. Smith, Nature New Biology *233*, 35 (1971).
(11)   Lindahl, T., and J. R. Fresco, in Methods in Enzymology, ed. L. Grossman and
       K. Moldave, Vol XII, part A, p. 601, 1967.
(12)   Richards, E. G. and W. B. Gratzer, Nature *204*, 878 (1964).
(13)   Loening, U. E., Biochem. J. *102*, 251 (1967).
(14)   Philipps, G. R., Analyt. Biochem. *44*, 345 (1971).
(15)   Philipps, G. R., and J. L. Timko, Analyt. Biochem., *45*, 319, 1972.
(16)   Aubert, M., J. F. Scott, M. Reynier and R. Monier, Proc. Nat. Acad. Sci., U.S., *61*,
       292 (1968).
(17)   Miura, K. M., Progr. Nucl. Acid Res. Mol. Biol. *6*, 39 (1967).
(18)   Lindahl, T., A. Adams and J. R. Fresco, Proc. Nat. Acad. Sci., U.S. *55*, 941 (1966).
(19)   Gartland, W. J. and N. Sueoka, Proc. Nat. Acad. Sci., U.S. *55*, 948 (1966).

Chapter 9. Clinical Applications

9.1 History and Future of Isozyme Separation Techniques

Robert L. Hunter

In our paper in which we introduced and defined the term "zymogram" we used three histochemical enzyme procedures, namely tyrosinase, alkaline phosphatase and nonspecific esterase, to illustrate the practicability of the application of histochemical enzyme methods following separation of proteins in starch gel. Perhaps because my principle academic responsibility was as a member of the faculty of a medical school it was my prediction at that time that the principle users of this new technology would be the clinical pathologists as they sought to more precisely define the enzyme protein changes which accompany health and disease in human beings. I expected that it would be only a few short years until clinical laboratories would be regularly utilizing these techniques. Actually this did not happen. The use of lactic dehydrogenase isoenzyme changes following heart attacks was clinically useful and did serve as an early example of the potential usefulness of zymogram technology. There are, of course, other examples of the use of isoenzyme analysis in support of clinical medicine but as one overviews the use made of this technique during the last 15 years it is soon obvious that it was not clinical medicine which put the technique to work. It was instead the cell biologists and the geneticists who immediately recognized the method's potential and today are utilizing in most elegant fashion the current refinements of zymogram technology.

One must ask why it was that these methods have not been more extensively used and several factors come immediately to mind. In the first instance, in the early days the technique of starch gel commenced with the purchase of a bag of potato starch from a grocery store and was followed by complicated chemical techniques to get an appropriate gel. Soon thereafter there was introduced a commercially available starch source which made the process much easier and the technology was widely adopted in research laboratories although clinical laboratories still preferred to use the lower resolution techniques. The existence of an extensive clinical literature based on observations utilizing low resolution electrophoretic techniques also worked against the adoption of the high resolution techniques. Another contributing factor was that the first widely used clinical laboratory method using isoenzymes, lactic dehydrogenase, did not require the high resolution technology; so there was little impetus to push ahead with high resolution electropherograms or zymograms.

The recent introduction of increasingly sophisticated automated techniques for the analysis of serum biochemistry suggests that technologies for understanding changes in multiple molecular forms will be developed once the scientific basis of these changes has been established and made important in terms of understanding human disease processes. Haemoglobin, although not an enzyme, is an example of a protein occurring in multiple molecular forms which have been extensively studied using electrophoretic techniques to the benefit of medical science. As we understand and study the multiple molecular forms of enzymes that compose serum we will be in an increasingly better position to diagnose and follow the prognosis of disease in conditions which are now escaping our attention. Most scientists are reluctant to simply go on "fishing expeditions" to see what changes occur in zymograms in disease states without having some preconceived intelligence as to why changes should occur. Because of this, much of the basic science necessary to forming the impetus for further utilization of these technologies is still lacking. Until such time as an easy to use, clinical electrophoretic technique is available in clinical-pathological laboratories I foresee a slow adoption of these techniques. Once they are appreciated in a few important circumstances the technology will be introduced and with its introduction will come utilization in an ever expanding appreciation of the potential of the technology. The symposium that we enjoy today here in Tübingen Germany, will serve as an important additional contribution toward the encouragement of the utilization of these techniques to the benefit of mankind. I wish this symposium every success.

9.2 Potential Applications of Polyacrylamide Gel Electrophoresis for Clinical Diagnosis

H. Hoffmeister

In determining the feasibility of the application of PAGE in clinical diagnosis means at the same time looking for proteins capable of giving information on defined conditions and courses of disease, influence of therapy etc. A selection of human serum proteins having such characteristics is shown in table 1. For a number of serum proteins, the biological function has not yet been clarified. However, all proteins described here have shown disease-specific behaviour (1, 2, 3). Of course, there are far more proteins used for diagnostic purposes (serum, blood, tissue enzymes; proteins in the urine and cerebrospinal fluid; proteins of defined tissues; other plasma and serum proteins).

Methods of immuno precipitation have been almost exclusively used to measure
alterations of serum levels in cases of disease so far (1,4). All those who are
familiar with these methods, know of the difficulties of standardization and
of varying influences on antisera due to storage and ageing. PAGE is an excellent
means of quantitative and qualitative characterization of the most important
serum proteins. Although the method is not a simple one, it presents a number
of advantages over immuno-precipitation: Pherograms show all proteins in
their interrelationships. The experienced worker is offered a characteristic
pattern indicating changes caused by disease. In quantitative determinations of
protein fractions, the methodological error will be small if a standard electrophoretic procedure is followed, staining and de-staining periods are exactly repeated,
and a uniform purified stain is used (5). In table 2, precision values in a series
of three peaks from our pherograms are reproduced.

Table 2. Precision in a series, demonstrated by 4 protein peaks from human serum
of genetic type Hp 2—1/Gc 2—1. The median values are based upon a calibration
curve established for human albumin in polyacrylamide gels (5).

Fraction	n	$\bar{x}$	s	CV
		g/l	g/l	%
Transferrin	34	13,84	0,82	5,9
Gc-Globulin/a_2-HS-Glycoprotein	33	5,74	0,44	7,7
Haptoglobin-2-2, Tetramer Fraction	36	5,44	0,28	5,2
β-Glycoproteins	33	1,14	0,14	12,4

However, diagnostic statements would presuppose that *normal pictures* of
serum electrophoresis for healthy persons were known, i.e. a definition of the
normal ranges for all protein peaks.
We have been using a strictly standardized and unaltered electrophoretic technique
with experimentally determined normal ranges for a number of years. To obtain
a uniform distribution of the important serum proteins over the whole length
of the gel, we have been using a *narrow* and a *wide* section of gel. This arrangement has been used for serum electrophoresis about 10,000 times. Although
a better separation of sera is achieved in the gel gradient or by the use of various
gel layers, we have not changed our system for reasons of comparability and
simplified operation. The normal values given in table 3 were derived from serum
protein pherograms for several hundred *healthy* persons.

Figure 1 demonstrates pherograms of gels from three normal sera which had been
studied by electrophoresis, stained, and de-stained under standard conditions.
These sera are of the three genetic haptoglobin types, Hp 2—2, Hp 2—1, Hp 1—1.
To evaluate disease-associated changes of the individual peaks, a confirmation is

Table 1. Human serum proteins of quantitative importance which can be determined with the aid of polyacrylamide gel electrophoresis. Their diagnostic importance has been revealed by immunological techniques and, in part, also by PAGE (1–4).

Protein	Biological Function	Importance in Diagnosis
Acid α_1-Glycoprotein	?	Increase : Neoplasms, PCP, Inflammations Decrease: Liver Diseases
Prealbumin	Thyroxine-binding Globulin, Retinol-binding Globulin	Decrease: Liver Diseases
Albumin	Osmot. Function, Protein pool, Binding of Ions, Dyes etc.	Decrease: Kidney Diseases, Neoplasms, Liver Diseases, Chron. Inflammations
α_1-Antitrypsin	Proteinase-Inhibitor	Increase : Inflammations Decrease: Liver Diseases
Gc-Globulin	Three Genetic Types	Decrease: Liver Diseases
Ceruloplasmin	Copper-binding Globulin, Oxidase	Increase : Pregnancy, Neoplasms Decrease: Morbus Wilson
Hemopexin	Hämin-binding	Decrease: Hämolytic Anaemias
Transferrin	Iron-binding	Decrease: Neoplasms, Inflammations, Paraproteinaemias, Nephrosis

Table 1, continued.

β_1 A-Globulin	Komplement-Factor	Increase : Infectiones, Toxoplasmosis Decrease: Autoimmun Diseases
Haptoglobins	Binding of Hämoglobin Three Genetic Types	Increase : Inflammations, Neiplasms, Infectiones, PCP, Nephrosis Decrease: Liver Diseases
α_2-Macroglobulin	Proteinase-Inhibitor	Increase : Liver Diseases, Diabetes, Nephrosis
α_1-Lipoprotein	Transportation of Lipids	Increase : Lipid Metabolism Disturbances Decrease: Liver Diseases
β-Lipoprotein	Transportation of Lipids	Increase : Nephrosis, Lipid Metabolism Disturbances Decrease: Liver Diseases
γA-Globulin	Antibody	Increase : Liver Diseases, Chron. Infectiones, PCP Decrease: Paraproteinaemias, Antibody Deficiency Syndrom
γG-Globulin	Antibody	Increase : Liver Diseases, Chron. Infectiones, Myelom, PCP Decrease: Paraproteinaemias, Antibody Deficiency Syndrom
γM-Globulin	Antibody	Increase : Chron. Infectiones, Macrogloblinaemia Waldenström, Liver Diseases

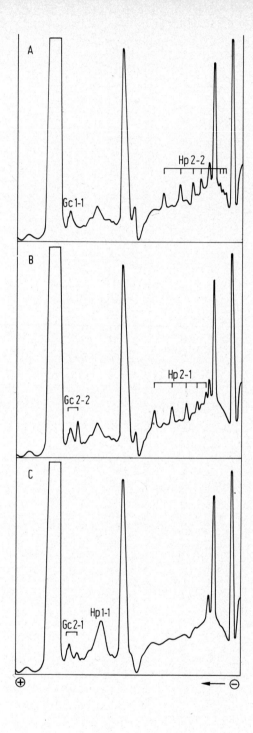

Table 3. Median values and normal ranges (± 2 s) of a number of serum proteins sampled from a group of probably healthy blood donors. The 2 s limits are corresponding to the ranges between the 2.5 and 97.5 percentiles assuming a normal distribution of values.

					Reference Range	
					$\bar{x} - 2s$	$\bar{x} + 2s$
Fraction	n	$\bar{x}$	s	CV		
		g/l	g/l	%	g/l	g/l
Albumin/a_1-Antitrypsin	60	47,8	0,92	1,9	46,0	49,6
Transferrin	312	2,22	0,22	9,7	1,78	2,66
a_2-Makroglobulin	312	0,52	0,15	28,5	0,22	0,82
Gc-Globulin/a_2-HS-Glycoprotein	120	1,67	0,37	22,2	0,93	2,41
β_2-Glycoproteins	312	0,43	0,095	22,2	0,24	0,62
γ-Globulins/β_1 A-Globulin	312	4,10	1,04	25,4	2.02	6,18
Hp 1–1, Monomer	33	1,60	0,46	28,0	0,68	2,52
Hp 2–1, 1. Polymer Fraction	147	0,13	0,048	38,0	0,03	0,23
Hp 2–1, 2. Polymer Fraction	147	0,12	0,028	24,0	0,06	0,18
Hp 2–1, 3. Polymer Fraction	147	0,074	0,019	25,5	0,04	0,11
Hp 2–2, 1. Polymer Fraction	132	0,14	0,044	30,6	0,05	0,23
Hp 2–2, 2. Polymer Fraction	132	0,15	0,042	28,2	0,07	0,23
Hp 2–2, 3. Polymer Fraction	132	0,10	0,028	27,5	0,04	0,16

needed to exclude changes due to genetic variants. Thus, the genetic type of a serum has to be accounted for in evaluation. There are also three genetic variants of Gc proteins which may be easily recognized in the pherogram (5); otherwise, a determination of these Gc types (Gc 1–1, Gc 2–1, Gc 2–2) will be difficult. Figure 2 schematically points out the area of the Gc type proteins. By standardized PAGE, a reliable classification of a serum to one of the three Hp and one of the three Gc types is possible, thus offering valuable information for forensic purposes.

Figure 3 shows those proteins which can be recognized and determined quantitatively or at least semi-quantitatively from PAGE. For comparison, the distribution of the proteins in cellulose acetate electrophoresis is given. Apart from albumin, distinct individual peaks are formed by acid a_1-glycoprotein, prealbumin,

---

Fig. 1. PAGE pherograms of three human sera of the genetic types, Hp 2–2/Gc 2–2 (A). Hp 2–1/Gc 1–1 (B), and Hp 1–1/Gc 1–1 (C). The lower 35 mm portions of the gels used for separation have narrow pores (7.5 % acrylamide; 1,6 % N,N'methylenbisacrylamide), the upper ones (40 mm) have wide pores (7.5 % acrylamide; 0.8 % N,N'methylenbisacrylamide). The gels stained with amido black have been evaluated in an Eppendorf photometer with pherogram accessory No. 2606, 0,25 mm aperture, at 578 nm.

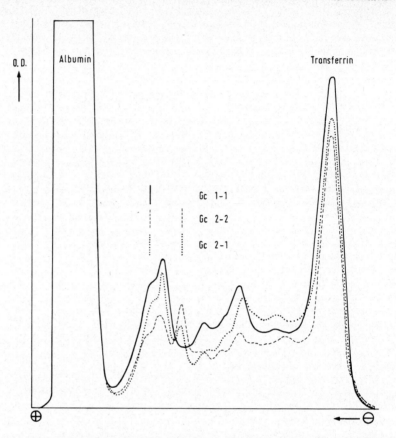

Fig. 2. Characteristic band pattern of the three genetically determined Gc protein types
in PAGE pherograms. In addition to Gc proteins, the group of bands also reveals a_1-anti-
chymotrypsin and a_2-HS-glycoprotein. Despite overlapping, the genetic Gc pattern is well
recognizable.

transferrin, β_1 A-globulin, 4—8 haptoglobin polymers, β_2-glycoprotein, a_2-macro-
globulin, and β-lipoprotein. There are also a number of other proteins which are
insufficiently separated like albumin and trypsin inhibitor; coeruloplasmin and
hemopexin; gc-globulin and a_1-HS-glycoprotein.

We have performed electrophoresis in sera from patients suffering from a variety
of diseases and in part of them, found typical deviations from the normal electrophoretic pictures. A schematic representation of a number of results is given
by figures 4—6. To facilitate the recognition of typical disease-associated changes,
each pherogram with pathological findings was compared to a normal pherogram,
for the same genetic Hp/Gc type. Part of the pathological findings has been

statistically confirmed already; all examples given are based on at least 50, and some on far more, cases that have been studied.

Figure 4 depicts a finding typical of a lung tumour, established on the basis of studies involving more than 90 confirmed cases of bronchial carcinoma (7). Even in cases with small tumours detected by chance, without subjective complaints, there was a highly significant increase of haptoglobins. Tumours in a progressive stage demonstrated an additional progressive decrease of transferrin and gamma globulin levels. Similar but not equally pronounced findings were derived from widespread inflammation, particularly in the area of the lungs. However, there was recognizable return to normal of haptoglobin peaks in cases of inflammation under the influence of therapy which fact is of importance for differential diagnosis. In some cases, haptoglobins were also influenced by tumours in other organs. However, we found in studies of tumours of the female genitalia, the stomach, the pancreas, and the intestine that the extent of influence and probalility of findings were less as compared with lung tumours.

Haptoglobins are particularly fast-reacting and sensitive serum proteins. This has been also demonstrated for primary chronic polyarthritis (PCP), the nephrotic syndrom, and toxoplasmosis. Pronounced pathological pictures of this type are characterized by extremely increased haptoglobin levels. However, in contrast ot neoplasms and chronic inflammations, transferrin will remain normal and the gamma globulins increase in the sera from PCP and toxoplasmosis patients. In addition, toxoplasmosis sera will show an increase of β_1 A-globulin. A pherogram typical for PCP is shown by figure 5.

Standardized PAGE may be used for the diagnosis of paraproteinemia with outstanding effectiveness. Figure 6 shows a pherogram of an atypical paraprotein in the γG range. When performing screening tests and examining suspect cases, we have repeatedly been able to demonstrate such small paraprotein peaks which in normal electrophoresis did not raise attention and were just recognizable with the aid of immuno electrophoresis. In our opinion, PAGE if applied by a skilled person is the most reliable diagnostic instrument for an early detection of paraproteinemia. The distinct protein peaks will also reveal whether monoclonic or polyclonic paraproteins are involved.

If concentrated, urine and cerebrospinal fluid will also yield pherograms which allow of a good evaluation by our system. Urine proteins will occur as individual peaks; most of them may be assigned to the corresponding serum proteins.

We have applied Freeman's method of immunological cross electrophoresis (8) to PAGE and obtained informative bidimensional pherograms (figure 7). The multitude of proteins which are present in serum becomes evident from the picture. It will be a practical impossibility to separate them by gel electrophoresis. In such a case, the gel would have to show a few hundred individual bands of

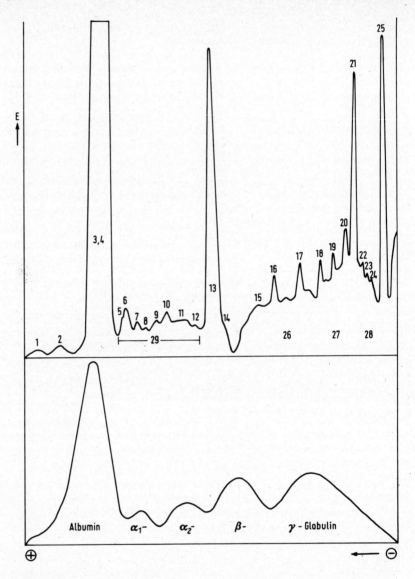

Fig. 3. Position of the most important serum proteins in PAGE. For the assignation of bands, various immunological standard techniques and partially also cross electrophoresis on cellulose acetate film (6) were used.

1 = prealbumin
2 = acid a_1-glycoprotein
3 = albumin
4 = a_1-antitrypsin
5, 7 = Gc-globulin

6 = a_1-HS-glycoprotein
      a_1-anti chymotrypsin
8 = ?
9 = ?
10 = coeruloplasmin

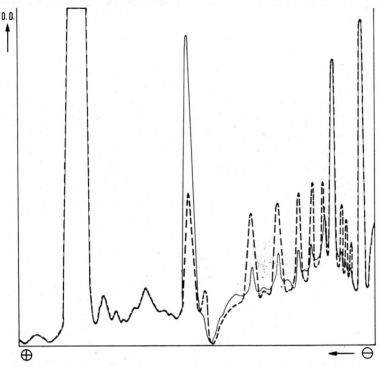

Fig. 4. PAGE pherogram typical of bronchial carcinoma. Hp peaks are considerably superelevated, transferrin and gamma-globulins reduced (fat line). For comparison a normal pherogram of identical Hp/Gc type is depicted (thin line).

11 = ?
12 = hemopexin
13 = transferrin
14 = inter-a trypsin inhibitor
15 = β_1 A-globulin
16
17
18 = haptoglobin polymers, genetic type 2−2
19
20 = β-glycoprotein
21 = a_2-macroglobulin
22
23 = haptoglobin polymers, genetic type 2−2
24
25 = β-lipoprotein
26 = γA-globulin
27 = γG-globulin
28 = γM-globulin
29 = a_1-lipoprotein

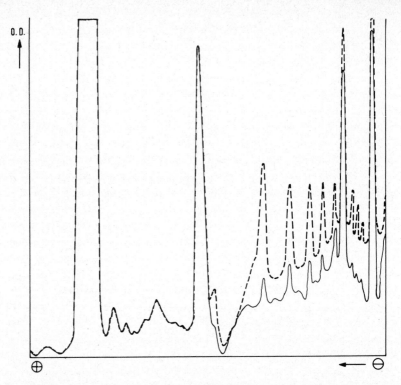

Fig. 5. PAGE pherogram of a serum derived from a case of primary chronic polyarthritis (fat line). A considerable increase of haptoglobins and gamma globulins is noted while transferrin and other globulins do not show recognizable changes as compared with a normal pherogram.

varying size. For diagnostic purposes, far more importance should be attributed to a utilization of the present achievement, namely serum protein separation by means of PAGE, under strictly standardized conditions and quality control. Cross electrophoresis is particularly suitable to give further information on the molecular uniformity of proteins. A number of precipitation lines are discernible, indicating the presence of identical antigenic characteristics in proteins which appear as more or less separate fractions in PAGE. This can be seen for instance in the case of haptoglobins. Figure 7 clearly shows precipitation lines with a number of peaks formed by Hp 2—1 serum. A similar pattern is shown by Hp 2—2 serum while one peak only is formed by Hp 1—1 serum. In cross electrophoresis, the patterns due to the three genetic Hp types will fully confirm the Hp peak patterns in PAGE as described by us earlier (9). It has also been possible to obtain the band patterns of the three genetic Gc types from anti-Gc serum. This phenomenon, however, is occuring in a considerable number of

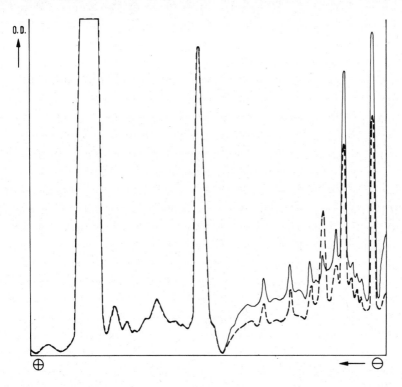

Fig. 6. Pherogram showing a small paraprotein peak in the γ G immuno globulin range.

other serum proteins. A particularly interesting one is the precipitation line obtained from anti-a_1-lipoprotein serum which frequently exhibits four maxima, thus suggesting the presence of at least four different a_1-lipoproteins in serum.

References

(1)   Becker, W.,W. Rapp, H. G. Schwick und K. Störiko..Z. klin. Chemie und klin. Bio-
       chemie 6, 113 (1968).
(2)   Hitzig, W. H., Die Plasmaproteine in der klin. Medizin, Springer Verlag, Berlin-
       Göttingen-Heidelberg 1963.
(3)   Hoffmeister, H., in Methodische Fortschritte im Med. Labor — Proteine, Proteide,
       Verlag Chemie Weinheim. In print.
(4)   Backhausz, R., Immundiffusion und Immunelektrophorese, VEB Fischer Verlag,
       Jena 1967.
(5)   Abraham, K., K. Schütt, I. Müller und H. Hoffmeister, Z. klin. Chemie und klin.
       Biochemie 8, 92 (1970).

(6) Hoffmeister, H., und K. Möller. Unpublished data.
(7) Schütt, K. H., und H. Hoffmeister, Z. klin. Chemie und klin. Biochemie 9, 201 (1971).
(8) Clarke, H. C., and T. Freeman, Protides of the biological fluids. Elsevier, Amsterdam 1966, Vol. 14, 503.
(9) Hoffmeister, H., K. Abraham, I. Müller und K. H. Schütt, 7th Int. Congr. Clin.-Chemie, Genf/Evian 1969 S. Karger, Basel, Vol. 1, 152 (1970).

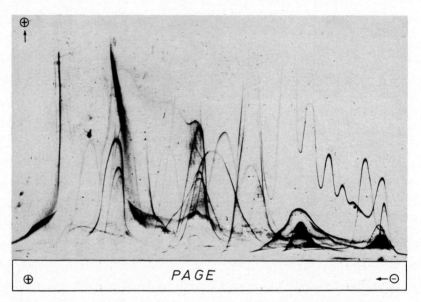

Fig. 7. PAGE picture followed by cross electrophoresis on cellulose acetate of genetic type Hp 2–1/Gc 1–1 serum against polyvalent equine anti-human serum. Many precipitation lines are exhibiting several peaks.

9.3 Changes of Serum Protein Patterns Associated with Liver and Kidney Diseases

K. Abraham[1] and *I. Abraham-Müller*[1]

In a preceding communication, Hoffmeister demonstrated the separation of the three different genetic haptoglobin-types using a method developed by us (1). Our aim consisted in separating electrophoretically the serum proteins of large numbers of patients with various diseases by a simple method practicable for

---

[1]   part of dissertation.

use in a routine laboratory; thus standardizing them should permit us to make precise comparisons.

To begin, I would like to show you the pherogram which derives from electrophoretically separating serum proteins of a HAA positive viral-hepatitis, which in turn stems from a series of diseases we have examined (figure 1). In this case, like in many others, we found striking changes in the haptoglobins (2,3). The monomer and the dimer are completely absent; the tetramer is highly reduced. At later stages of the disease the tetramer is absent and the octamer is reduced.

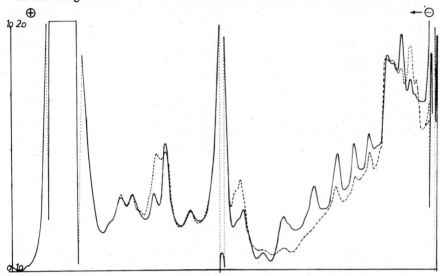

Fig. 1. HAA-pos. viral-hepatitis; the long straight line denotes the appearing of the first clinical symptoms, the dotted line refers to their appearance two weeks later. 10 nl, amido black 10 B, evaluation at 546 nm in an Eppendorf photometer.

It is necessary to point out that corresponding to the seriousness and the stage of the disease there may occur a great variety of transitional states. Certainly, the absence of the haptoglobin-polymers is not dur to a lack of synthesis of haptoglobin. If one incubates the serum of normal persons at ph 5 with enzymes taken from the liver, at increasing concentrations, one obtains patterns which are identical to those found in the serum of viral-hepatitis patients; in which the concentration of monomer, monomer-dimer, monomer-dimer-tetramer, etc. are reduced or absent. Since necrosis of liver cells is a concomitant to viral-hepatitis, (4) and since enzymes are released (5,6), we feel we can explain this pattern of haptoglobin. If we divide the soluble cytoplasmatic proteins of liver-autopsy-cylinders, there is no evidence for proteins, (7,8) whereas in liver-autopsy-cylinders of normal persons we find a specific pattern. This finding is confirmatory to the results obtained in histological sections.

In addition to changes defined as liver-specific, there are changes in the region of the β-glycoprotein and a_2-macroglobulin. Although the β-glycoprotein is indistinct, two bands can be resolved. The a_2-macroglobulin is reduced considerably and in the direction of the cathode sometimes one, sometimes two distinct bands can be seen. We detected this pattern in other viral-infections, such as influenza, ornithosis, and psittarcosis, which are marked additionally by an increase in complement (14).

Following the incubation of normal sera with lysozyme we obtained the same type pattern. High concentrations of lysozyme occur above all in the granules of polymorphonuclear leucozytes and in lungs-macrophages (9,10). Even though viral-infections are accompanied by leukopenia, and even though influenza, ornithosis and psittacosis are accompanied by pathologic changes in the lungs, further investigations are necessary to determine if these changes are due solely to lysozyme.

Another parameter consists in the increase of immunoglobulins (15), that is to say especially in the IgM which can be localized in the region between a_2-macroglobulin and β-lipoprotein in our gel-system. The three changes described above are consistent with a diagnosis of viral-hepatitis. In addition to this, we can judge the degree of seriousness of the disease by the pattern of haptoglobins and by the immunoglobulins (for example: no evidence of haptoglobins), or we can diagnose the persistence of hepatitis (for example: continuing absence monomer and dimer etc. for periods of months).

The potential diagnostic value of these techniques in alcoholic liver cirrhosis is shown in figure 2. From inflammatory fatty liver (alcoholic hepatitis) to alcoholic liver cirrhosis the concentration of IgG and especially IgA increases. In comparison to normal sera one finds a distinct increase of IgA and IgG(11, 12, 13) see figure 2. Haptoglobin polymers are not represented, and we have not done a haptoglobin-type 1–1. The rest of the haptoglobin-types can be shown by adding hemoglobins to the sample (figure 2 dotted line). In advanced stages like this the haptoglobin-type is not at all, or only partially definable and the transferrin varies from low to low normal levels. Usually, it is not as easy to diagnose liver cirrhosis, since the different types of cirrhosis cannot be delineated by statistical evaluation of the patterns alone. Another reason for this lack of clear delineation stems from the transitional state between chronic hepatitis and frank liver cirrhosis.

These difficulties of differentiation orginate mainly in histology which provides diagnoses of a highly differentiated nature. We must ask therefore, whether histology actually reflects the state of the whole liver in any given case. Unfortunately, our investigations in this field are not sufficiently developed to permit us to make definitive diagnoses as yet.

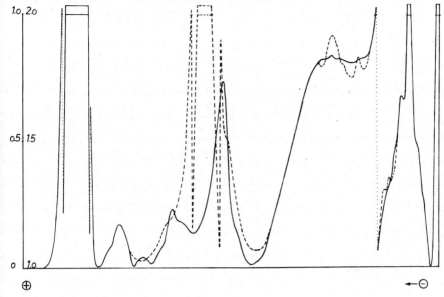

Fig. 2. Alcoholic liver cirrhosis; rests of haptoglobin can be shown by the loading with hemoglobin, dotted line. 10 nl, Amido black 10 B, evaluation at 546 nm in an Eppendorf photometer.

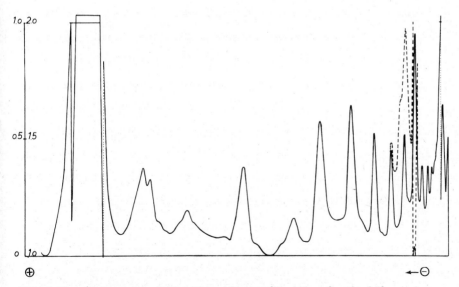

Fig. 3. Chronic glomerulonephritis with impairment of the kidney function before and after (dotted line) dialysis. 10 nl, Amido black 10 B evaluation at 546 nm in an Eppendorf Photometer.

PAGE separation of the serum-proteins of patients suffering chronic glomerulo-nephritis, results in a distinct increase of haptoglobins, too (figure 3). While with liver-diseases we find a loss of haptoglobins, here the haptoglobins increase two or three times above normal values. Especially remarkable is the increase of H in the Hp 2−2 and the Hp 2−1 type. This increase of haptoglobins is not due to the overloading with hemoglobin, as could be demonstrated by the benzidine-reaction. The immunoglobulins IgA and IgG are deminished as complement C_3. Patients who suffer from a chronic kidney insufficiency in connection with chronic glomerulonephritis and have been dialysed demonstrated changes in the region of the a_2-macroglobulin after dialysis (figure 3, dotted line). Since on the one hand the patients receive heparin before each dialysis, and since on the other hand, the bands of fibrinogen are situated at this point, we can possibly attribute these changes to fibrinogen. They disappear within 24 hours.

Besides a continuous slight increase of the haptoglobins and a decrease of trans-ferrin, no striking displacement in the protein pattern during dialysis can be found. Following, however, the protein-pattern of dialysed patients for months, we were able to detect a decrease of haptoglobins to the point of complete disappearance. This means not that these changes depend on the creatinin-or urea-concentration of the serum, even though it is possible to destroy the hapto-globins by incubation with high urea-concentrations in experiment. Certainly we cannot attribute these changes to the kidney, but, as a high percentage of dialysed patients suffer chronic hepatitis, I think it is possible or even probable to consider this phenomenon as an acute push.

In conclusion, I suggest that the statistical significance of chronic glomerulone-phritis with impairment of the kidney-function and its differentiation against normal serum patterns of persons, liver-diseases, and inflammatory lungs-diseases can be regarded as diagnostically successful, whereas the task of obtaining meaningful statistics, for example, of the lung, has not yet be completed.

References

(1) Abraham, K., K-H. Schütt, I. Müller und H. Hoffmeister, Z. klin. Chem. u. klin. Biochem. *8*, 92 (1970).
(2) Schütt, K.-H. und H. Hoffmeister; Z. klin. Chem. u. klin. Biochem. *9*, 201 (1971).
(3) Hoffmeister, H., K. Abraham, I. Müller und K.-H. Schütt, VII. Intern. Congress of Clin. Chem., Genf 1969, Kongreßberichte Bd. *1*, 152.
(4) Popper, H. und F. Schaffner, Die Leber, G. Thieme Verlag 1961
(5) Schmidt, E., 2. Konferenz d. Gesellschaft für Biologische Chemie, Enzymdiagnostik, 1967.
(6) Gerlach, M., 2. Konferenz der Gesellschaft für Bibl. Chemie, Enzymdiagnostik, 1967.

(7) London, M., R. Mc Hugh and P. B. Hudson, Cancer Res. *14*, 718 (1954).
(8) Henley, K. S., H, S. Wiggins, H. M. Pollard and E. Dullaert, Gastroenterology *36*, 1 (1959).
(9) Humphrey, J. und R. G. White, Immunologie, Georg Thieme Verlag 31 (1971).
(10) Salton, M. R. J, Bacteriol. Rev. *21*, 82 (1957).
(11) Scheurlen, P. G., Der Internist *9*, 56 (1968).
(12) Fateh-Moghadam, A., R. Lamerz, J. Eisenburg und M. Knedel, Klin. Wochenschrift *47*, 129 (1969).
(13) Gleichmann, E. und H. Deicher, Klin. Wochenschrift *46*, 793 (1968).
(14) Schütt, K.-H., pers. report
(15) Gleichmann, E., H. Deicher, Klin. Wochenschrift *46*, 171 (1968).

9.4 Comparative Disc Electrophoresis of Serum and Cerebrospinal Fluid

K. Felgenhauer

The question is pertinent, whether polyacrylamide gel electrophoresis has proven to be as fruitful and useful a method for the clinical laboratory, as it has developed for the biochemical laboratory. Eight years have passed, since the first symposium about gel electrophoresis in New York (1), many technical modifications have been developed since (survey: 2) and it is not premature to consider its value for clinical chemistry at present.

Of course the answer will depend on the criteria for success one is willing to apply. Is there any diagnostic problem, which can be resolved only with polyacrylamide gel electrophoresis? Has it replaced one of the routine electrophoretic screening techniques or broadened at least their application?

Two techniques have to be compared in this respect, free zone electrophoresis and immunoelectrophoresis. The first, especially if properly performed in agarose is now a very useful screening method for its technical simplicity and easy interpretation. It is mainly used to answer the following questions:

1. *Are there quantitative immunoglobulin abnormalities?*
   The gel techniques so far have not been able to give a better answer to this question than free zone electrophoresis. Immunoglobulins are intermingled with Haptoglobinpolymers, a_2-Macroglobulin, β-Lipoprotein, Plasminogen and $\beta_{1A/C}$ Globulin. Even if fast densitometry is possible there is no meaningful way and actually no sense to substract all these proteinbands to obtain values for immunoglobulin quantitation.

2. *Are monoclonal immunoglobulins present?*

It needs considerable experience to detect unequivocally abnormal proteins
in the immunoglobulin region. This is especially then the case, if the paraprotein concentration is in the same range as the neighboring normal proteins
or if the patient belongs to an unusual haptoglobin type. It is certainly no
problem to detect a paraprotein, present in high concentrations, especially,
if the patient belongs to the haptoglobin type 1—1.

There is, however, a much more serious obstacle in using the Ornstein-Davis
discontinuous technique for γ-globulins. It can be shown by postelectro-
phoretic immunoprecipitation, that certain immunoglobulin fractions do not
penetrate the separation gel and are retarded in the turbid stacking gel, which
is generally thrown away. Therefore a myeloma protein may be missed and
this is an intolerable risk for the clinical chemist.

We have therefore tried a transparent upper gel, which can be stained and
recorded together with the lower gel (3). The running conditions of the Ornstein-Davis system were essentially preserved in this gel, but chloride was
used instead of phosphate as leading ion and photopolymerization was replaced
by catalyst polymerization. Acrylamide concentration was reduced from
3.5 to 2.8% and the degree of crosslinking was considerably lower. The lower
gel was not changed.

On application of this technique in comparison to conventional techniques
now all γA-and those γG-paraproteins with isoelectric points below 8.5
became visible (4).

Since very low paraprotein concentrations are detectable, this technique may
be used in cases with clinical myeloma suspicion and a normal immunoelectrophoretic pattern. We have found one patient suffering from an extradural
and later generalized myeloma, who had a normal immunoelectrophoretic
pattern, but a distinct monoclonal band in serum and CSF (figure 1). But
even with a transparent upper gel several γM-paraproteins and some γG-paraproteins with isoelectric points above 8.5 still do not migrate into the transparent concentration gel.

3. *Are there a-globulin abnormalities?*

Since the a-globulins differ widely in molecular size, they are optimally separated on gels. They now are distributed over the whole pattern, but fortunately most of them are glycoproteins and can therefore selectively be
stained with the PAS method. But the PAS technique has not been tried as
a screening technique for a-globulins. The Amidoblack pattern on the other
side is very complex and variable, making its interpretation almost impossible in the routine clinical laboratory, at least at the present time.

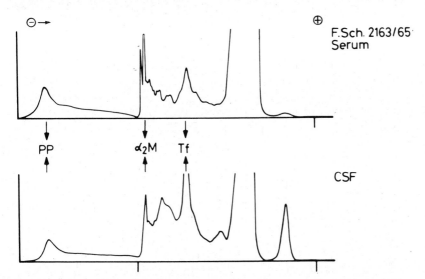

Fig. 1. Disc electrophoretic patterns of serum and cerebrospinal fluid in a 47 years old patient, suffering from multiple myeloma. Five months prior to electrophoresis an extradural tumor of the spine was removed. Histologic examination: Myeloma.
The immunelectrophoretic serum pattern was normal, a Bence Jones protein was excreted with the urin.
CSF: 29 mg % protein; 2 cells per mm^3; colloidal mastix reaction: 1, 3, 5, 5, 3, 1, 1, 1, 1, 1,
Abbreviations: a_2M = a_2-Macroglobulin, PP = Paraprotein, Tf = Transferrin.

Polyacrylamide gel electrophoresis has proven to be a very valuable tool to evaluate protein filtration through biological membranes. Restricted diffusion is the dominating basic principle of this process. Diffusion velocities of proteins in membranes decrease considerably stronger with increasing molecular size than in free diffusion (5). Filtrates therefore contain large amounts of low molecular proteins amd small amounts of large proteins, the ratio depending on the membrane properties. Since the loading capacity of disc electrophoresis is high and band spreading very low, especially with the larger components, the technique is optimal suited to compare the proportions between small and large proteins before and after filtration. Most biological filtrates originate from the serum like glomerular filtrate, eye chamber fluid, amnion fluid, lymph fluid and cerebrospinal fluid. The serum bands compiled in table 1 (R_{ES} = Einstein-Stokes-Radius) are easy to identify in 6–7 % gels and may therefore serve as convenient parameters for this purpose.

Transferrin is easy to identify, but intermingled with hemopexin. The proteins of the postalbumin region are very complex, the immunoglobulins do not form dominant peaks and Plasminogen is not clearly separated from β_{1A}- and β_{1C}-globulin.

Table 1

	MW	$R_{ES}(\text{Å})$
Prealbumin	61 000	33
Albumin	69 000	36
Hp 1−1	80 000	44
Hp-Polymers	> 87 000	> 54
a_2-Macroglobulin	798 000	94
β-Lipoprotein	2 239 000	124

The patterns of serum and cerebrospinal fluid have been compared in normal and pathological conditions (6). Comparing the normal patterns the following features are characteristic:

1. The small prealbumin is present in comparable large amounts.
2. Haptoglobin 1−1 is present in somewhat lower relative amounts.
3. The third postalbumin and the second posttransferrin (= τ-Globulin) are CSF-typical bands.
4. The haptoglobin-polymers, mainly those of higher order, are indetectable.
5. a_2-Macroglobulin is always present in small amounts.
6. β-Lipoprotein is virtually absent.
7. The immunoglobulins of the transparent upper gel, mainly comprised of IgG, are considerably more weakly stained.

It became evident during these studies that for all disease states a clearcut decision was possible, whether the elevated protein level was caused by a disturbance of the blood -CSF-barrier, or by a synthesis of proteins within the central nervous system. The term blood-CSF-barrier defines the state of dynamic equilibrium between the intravascular compartment and the subarachnoidal space. Disturbances of that barrier will produce an approximation of the CSF to the serum pattern, characterized by the following alterations:

1. The highmolecular proteins increase strongly, e. g. the a_2-Macroglobulin, β-Lipoprotein and the haptoglobinpolymers.
2. The low molecular proteins like prealbumin constitute a smaller relative proportion of the total protein amount.
3. CSF-typical bands like the 3rd postalbumin become indetectable.

In most neurological diseases a barrier disturbance is the dominant cause for an abnormal cerebrospinal fluid composition, for instance in all acute inflammations, whether of bacterial or viral origin, but also in tumours and vascular diseases of the central nervous system.

Those diseases where new proteins, synthesized in the central nervous system become detectable deserve much more interest. To our present knowledge these proteins belong mainly to the Immunoglobulin class. The blood-CSF-barrier is intact and the distribution profile is generally different from that of the patients serum. This is generally the case in syphilitic diseases, multiple sclerosis and in subacute sclerosing leucencephalitis.

There is a third abnormal pattern prototype. A barrier disturbance is combined with a more or less strong local immunoglobulin production, as seen in late stages of bacterial meningitis, circumscribed myeloma infiltration or mycotic infections of the brain.

While gel techniques have not found their way into routine clinical chemistry until now, they have a high heuristic potential. In our case they have provided a sound theoretical basis for the interpretation of single protein determinations in CSF and a better understanding of the blood-CSF-barrier in normal and pathological conditions.

References

(1)   Gel Electrophoresis. Ed.: H. E. Whipple. Ann. N. Y. Acad. Sciences, *121*, 305–650 (1964).
(2)   Maurer, H. R., Dics Electrophoresis. W. de Gruyter, Berlin, New York 1971.
(3)   Felgenhauer, K., Clin. Chim. Acta *39*, 175–181 (1972).
(4)   Felgenhauer, K., G. Alzer and K. Schumacher, Klin. Wschr. 50, 1033–36 (1972).
(5)   Renkin, E. M., J. Gen. Physiol. *38*, 225–243 (1955).
(6)   Felgenhauer, K., Vergleichende Disc-Elektrophorese von Serum und Liquor cerebrospinalis. Georg Thieme, Stuttgart 1971.

9.5 Electrophoretic Separation of Pre-stained Lipoproteins on Polyacrylamide Gel Slabs and their Relationship to Other Plasma Proteins

R. C. Allen

The use of polyacrylamide gel electrophoresis (PAGE) for the separation of pre-stained lipoproteins offers a potentially rapid qualitative and quantitative method for the clinical study of the hyperlipoproteinemias. However, the method has not been extensively employed. Presumably its lack of general acceptance as a routine clinical tool lies in part in the greater technical difficulty in preparing reproducible gels and in the fact that until the recent work of Moran et al. (1) in PAGE that the beta lipoprotein required the use of ultracentrifugation for its accurate quantification.

The affects of various parameters, such as pre-stain preparation and gel concentration (2, 3), buffer ions (4) and sample handling; all play important roles in separation, resolution, and in the subsequent qualitative and quantitative accuracy of gel techniques interpretation. Raymond (2) has reported that gel monomer concentrations of greater than 3.75% inhibit the migration of the beta and presumably pre-beta fractions. While Klemens (5) has observed that the very low density lipoproteins (pre-beta) will not migrate in a 3.0 % gel if the triglyceride concentration exceeds 800 mg per 100 ml. On the other hand, the use of gels with pore sizes sufficiently large ot allow migration of the chylomicron and pre-beta (VLDL) and the LDL (beta) do not provide optimal sieving effects to adequately resolve HDL (High density alpha lipoproteins) and are exceedingly difficult to handle for subsequent densitometry. Thus, a simple step gradient gel appears to offer the best characteristics for optimal separation where a region of large pore size gel for VLDL and LDL separation is provided and a smaller pore size gel for optimal HDL resolution is also present. Such a system has been employed by Wollenweber and Kahlke (3) using a 3.25–7.5 percent T gel in a modified Ornstein (6) and Davis (7) method to compare PAGE and Agarose methods systems. In this system chylomicron remained at the origin and the prebeta migrated only a short distance into the 3.25 percent gel. No quantitative studies were attempted by these authors.

On the other hand, Moran et al. (1) compared the quantification of the beta-lipoprotein with a modified *Quick Disc* method (8) with that obtained by ultracentrifugation studies and demonstrated an excellent correlation of results.

Since step gradient gel slabs are readily and simply prepared, as previously described (9), we sought to develop such a system for the separation of lipoproteins which at the same time could be used to correlate lipoproteins with other plasma proteins and isozymes. Initial qualitative and quantitative studies on such a system are the subject of this report.

Methods

Step gradient gel slabs were prepared in vertical electrophoresis cells (ORTEC) utilizing a first step 10 mm in depth of 3 percent T with 3.5 percent C layered over an 8 percent T, 3.5 percent C gradient sufficient to fill the remainder of the cell. Additionally, 3–5–8 percent and 3–6–8–12 percent gradients were employed for correlative studies. The gel buffer employed was tris-citrate at pH 9.0. A 0.0375 M (final concentration) buffer was used in the 3 percent layer, well and cap gel, while 0.1875 M (final concentration) was used in the succeeding layer or layers as previously described (10). Recrystallized acrylamide and methylene bis-acrylamide were employed only for reasons previously reported (11).

Lipoprotein pre-staining solution: 24 milligrams of Sudan Black were dissolved in a mixture of 24.40 ml of ethylene glycol and 0.6 ml of 1.5 M triscitrate buffer pH 9.0. (Use of greater amounts of aqueous buffer greater than 15 % will cause the Sudan Black to precipitate out of solution.) The solution was warmed to 60°C for one hour and then filtered through Whatman number one filter paper or equivalent.

Esterases: Esterases were separated in 3–6–8–12 % T gradient gels, and the multiple molecular forms of the non-specific plasma esterases were determined utilizing alpha-naphthol acetate as the substrate and Fast-Red TR as the complexing agent.

Haptoglobulin typing: Haptoglobins were also determined in 3–6–8–12 % T gradient gels. Samples were mixed with hemoglobin from a 1–1 haptoglobin type, and following electrophoresis, were reacted with benzidine and barium peroxide tablets (Merk) in acetic acid 5 ml H_2O and 5 ml glacial acetic acid to develop the hemoglobin.

Sample preparation: Control plasmas were obtained from fresh citrated whole blood samples (0.1 ml of 3.8 % sodium citrate per ml of whole blood). VLDL, LDL and HDL lipoprotein preparations were kindly supplied by Dr. U. H. Klemens, Klinikum Steglitz, Berlin, and were prepared by ultracentrifugation as previously described by him (12). Clinical material was obtained as fresh serum from the University Clinics, University of Heidelberg, courtesy of Dr. Kahlke.

Fifty microliters of fresh plasma or serum were added to 50 microliters of the pre-stain solution in Eppendorf model 3200 centrifuge tubes. The mixture was shaken and then incubated for 30 minutes at 37° C. Following incubation, the samples were centrifuged for one minute in an Eppendorf model 3200 clinical centrifuge. Twenty microliters of the mixture (approximately 600 μgm protein as determined by Biuret) were layered onto the gels under a 4.0 % cap gel and the cap gel was then allowed to polymerize before inserting the cell into the tank.

Electrophoresis: Electrophoresis was carried out in an ORTEC model 4200 system utilizing an ORTEC 4100 pulsed constant power supply at 280 V for one cell and 330 V for two cells. The pulse rate was increased from 50 pulses per second at the start of the separation to 100 pulses per second at 5 minutes and then to 200 pulses per second at 15 minutes. Separations were carried out until the albumin had migrated a distance of 3.3 cm from the origin, 40–60 minutes depending on the gel gradients employed.

Microdensitometry: Microdensitometry of the gel slabs was performed with an ORTEC model 4300 integrating microdensitometer.

Results

Comparison of the lipoprotein patterns of plasma and serum pre-stained with
Sudan IV and similarly pre-stained VLDL, LDL, and HDL fractions obtained
by ultracentrifugation are shown in figure 1 a and 1 b. The pre-staining procedure
as well as post-staining with lipid crimson (Gurr) were compared and equivocal
qualitative results were obtained. However, densitometric quantification indicated
a marked background in the VLDL-LDL region and a lower dye binding affinity
for the HDL region post-stained with lipid crimson. Chylomicron, pre-beta,
floating beta and beta readily resolved in the 3 percent T, 3.5 percent C gel layer.
Best resolution of the alpha region was obtained between 8 and 10 % gel gradients
where there appear to be at least four-five distinct bands present in the HDL region

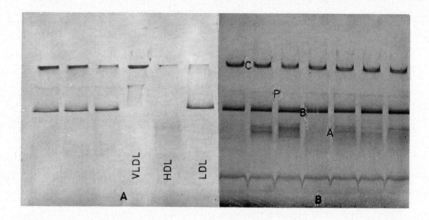

Fig. 1. A, Separation of serum and ultracentrifuge fractions of the serum lipoproteins
pre-stained with Sudan black. B, Separation of lipoproteins from fresh plasma, VLDL
chylomicron (C), pre beta (P), LDL beta (B), HDL alpha (A) indicating the various
HDL patterns obserbed. Separation in a 3–8 % T, 3.5 % C gel.

Quantitative Results

To preliminarily assess the quantification of lipoproteins pre-stained with Sudan
Black, pre-stained ultracentrifuge fractions of VLDL, LDL, and HDL were
separated by PAGE. Results shown in table 1. Indicated that, in the ranges
found in both normal plasma and serum and in that of hyperlipoproteinemias,
dye binding was quite linear. A small amount of stained material was found
to remain at the origin in both the LDL and HDL fractions (figure 1 a) which
may have been due to some denaturation during the shipment time between
the two laboratories.

Table 1. Quantitative microdensitometry of pre-stained ultracentrifuge fractions separated by PAGE.

Protein Applied	VLDL		LDL		HDL	
	200 μg	100 μg	37 μg	18.5 μg	730 μg	365 μg
Chylomicron region	53.88	29.02	6.16	2.90	15.88	7.38
Pre-beta region	7.34	3.72				
Beta region			50.18	23.48		
Alpha region					58.99	25.46

units expressed as integrator count x 10^{-3}

Quantitative reproducibility and the effect of storage of plasma on the quantification of the various lipoproteins is shown in table 2 where the HDL decreases with storage and the VLDL markedly increases.

In order to compare the potential of the classification of the hyperlipoproteinemias by quantitative PAGE with standard biochemical and Agarose techniques, normal levels of the plasma lipoproteins separated by PAGE were determined in a group of 18 normal individuals as shown in table 3. Values expressed as arbitrary units were obtained from the total integrator count of each lipoprotein peak multiplied by 10^{-3} and corrected to 600 ugm of total protein. Since the males in this study had significantly higher chylomicron levels ($P > 0.01$) than females, normal values were selected from this group for purposes of comparative classification. Thus, chylomicron levels of above 25 units, pre-beta above 4.5 units, and beta levels above 33 units, or over 2 standard deviations from the normal group, were considered as being elevated. A comparison of type classification by both quantitative PAGE based on the scheme of Fredrikson, et al. (13) and

Table 2. Quantitative reproducibility and storage effects on plasma lipoprotein values from a single individual.

Plasma	Protein	VLDL		LDL	HDL
		(chylomicron)	(pre-beta)	(beta)	(alpha)
Fresh 23 June	600 μg	9.50	2.50	30.00	37.60
Fresh 14 July	600 μg	14.22	3.11	33.70	33.20
Fresh 11 Aug.	600 μg	12.60	2.70	29.00	29.00
Fresh 14 Aug.	600 μg	12.50	2.75	29.20	32.00
Mean		12.10 ± 1.97	2.77 ± 0.25	30.48 ± 2.19	32.95 ± 3.56
Stored plasma from 11 Aug.					
2 days 4° C	600 μg	17.50	3.80	33.00	28.00
3 days 4° C	600 μg	22.38	5.34	34.40	27.00
7 days 4° C	600 μg	30.90	8.74	35.60	21.00

Table 3. Comparison of Hyperlipoproteinemia Classification by Agar Rose and PAGE

| Subjekt | Sex | Age | Tri-Glyceride | Cholesterol | Classification | | Quantification of PAGE* | | | |
					Agar Rose	PAGE	Chylomicron	Pre-Beta	Beta	Alpha
1	Male	37	1500	400	IV	V	32.41	7.65	17.81	25.91
2	Male	40	740	436	V	V	35.85	7.50	40.00	20.24
3	Male	70	750	328	IIb**	I	29.03	3.40	24.18	23.40
4	Male	39	135	290	IIa	IIa	18.60	2.70	33.30	46.99
5	Male	59	130	404	IIa	IIa	13.55	2.22	35.10	32.22
6	Male	37	205	286	IIb	IIb	21.7	6.80	37.18	45.60
7	Male	NA	200	252	N***	IIb	21.60	4.26	39.64	39.34
4	Males	23–60	–	–	N	N	19.53 ± 1.54	2.76 ± 0.83	26 ± 3.71	33.27 ± 4.36
14	Females	20–60	–	–	–	N	11.57 ± 3.09	1.99 ± 0.99	24.02 ± 5.04	46.14 ± 6.8

* Quantity expressed as integrator counts x 10^{-3} corrected to 600 μg protein

** Typical type V pattern on Agarose but with high B-cholesterol

*** Classified as a borderline normal on Agarose

Klemens (14) and by biochemical and agarose electrophoresis was made independently and the results of the two methods are summarized in table 3, while typical densitometric scans are shown in figure 2. Reasons for agreement or lack of agreement of the type classification by the two methods are discussed later.

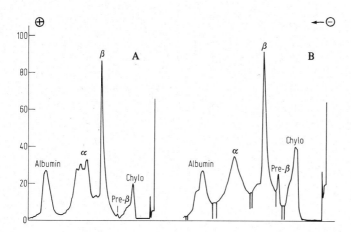

Fig. 2. Densitometric scan of normal plasma (A) and a type II b (B) stained with Sudan Black.

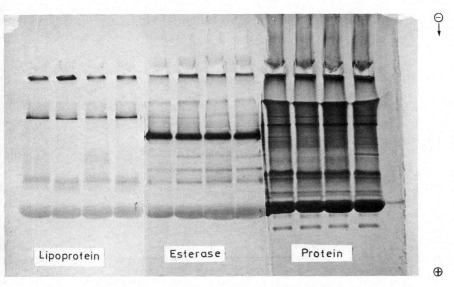

Fig. 3. Separation of four different Alpha lipoprotein types and their associated esterase and protein patterns. Separation in a 3–6–8–12 percent T, 3.5 percent C gel.

In figure 3, four different plasmas were separated from a 3—6—8—12 % T, 3.5 % C gel and stained for lipoprotein, esterase and protein actibity in order to assess the relationship of the lipoproteins with other plasma proteins and iso-zymes. Chylomicron, pre-beta and the beta lipoprotein fractions all demonstrated corresponding esterase activity and, as may be noted, sample three with a slow beta also shows the correspondingslow esterase. One of the HDL alpha bands also appears to have associated esterase activity. The most rapidly found band labeled 5 in figure 1 may correspond to Band A, HDL 3 types described by Uterman (15). It has been observed to be present in only seven out of some 50 samples tested and so far has been found associated only with 1—1 haptoglobin types.

Discussion

The results obtained indicate that the lipoproteins are readily resolved into chylomicron, pre-beta, beta, floating beta, and four or more alpha bands in the step gradient slab gel system described. Pre-staining with Sudan Black in buffered ethylene glycol appears to provide a rapid method for both subsequent qualitative and quantitative studies.

Quantitative densitometric studies on pre-stained VLDL, LDL, and HDL ultracentrifuge fractions indicated that over similar ranges that occur 600 ugm of human plasma or serum protein, that the dye binding is quite linear in the chylomicron region and the pre-beta, beta and alpha lipoproteins. The increased amount of Sudan Black stained material remaining at the origin in the HDL centrifuge fraction may have also been due to parial denaturation in the time required for shipment between the two laboratories. This is consistent with the quantitative decrease in HDL and increase in the chylomicron region found on storage of plasma and serum, however, the source of the small amount of material from LDL remaining at the origin is not clear. These results do suggest, however, that lipoprotein separations should be carried out on fresh plasma or serum, otherwise a true type IV familial hyperlipoproteinemia could possibly be classi-fied type V in a non-fresh sample or a II a could be confused with a II b.

The ability to clearly resolve and readily quantify the VLDL (pre-beta) and LDL (beta) by microdensitometrical methods following separation by PAGE appears to offer an advantage in distinguishing types II a and II b. Borderline cases such as numbers 3 and 9 in the table 2 are a good example of this problem. Serum from patient number 7 was classified biochemically and by agar gel electro-phoresis as a borderline normal, yet the triglyceride and cholesterol levels are similar to patient number 6 which was classified as II b by both methods. Quanti-tative PAGE data on the other hand indicates both have a high beta and pre-beta, thus suggesting that number 7 is a type II b, also.

The discrepancy in classification between the two methods in sample 1 as to whether it is type IV or V points out the necessity of utilizing fresh sample material since a type change between IV and V has been described by Klemens (14). This potential is also shown in table 3 where a decrease in HDL and increase in VLDL chylomicron and pre-beta occur on storage, and could thus easily shift a type IV to a type V with quantitative PAGE classification.

The classification of sample 3 as a IIb based on beta-cholesterol levels, but with a typical type 5 pattern on Agarose electrophoresis, is confusing. It appears by PAGE electrophoresis to be a type I since by PAGE both pre-beta and beta are not elevated while chylomicron is considerably higher than normal. Unfortunately, there was no opportunity to repeat this case with fresh plasma or serum. In this light, however, PAGE also offers the advantage of differentiating between the alpha Lp (a) (16) and the pre-beta which have the same mobility on agar gel, since the sieving effect of acylamide causes the pre-beta to be located cathodal to the beta rather than anodal. Thus, in electrophoresis in nonsieving media, confusion between IIa and IIb or a primary type IV pre-beta hyperlipoproteinemia could be falsely identified as a consequence of high Lp (a) levels.

Heretofore, in agar gel lipoprotein electrophoresis the quantitative alpha levels appear not to have been considered as major contributors to the type classification of the lipoprotein pattern. According to the artherosclerosis potential (AP) test performed with the *Quick Disc* method this region is reduced in both types I and V. This is consistent with the type I and V classifications from PAGE quantitative densitometric found in this preliminary study. However, further studies are required to establish whether or not the quantity of alpha lipoprotein can serve as a criterion for type differentiation. On the other hand, the significance of the various HDL lipoproteins remains to be elucidated. Loss of the more rapidly migrating ones is seen in pigs following trauma and/or endotoxin shock (17).

Initial studies to correlate the lipoproteins with corresponding proteins and isozymes indicated that in the chylomicron and pre-beta VLDL associated esterase activity was inversely proportional to the quantity of lipoprotein present. However, further work is required to assess the relationship of the esterase activity to the amount of lipoprotein and its possible significance, if any, in type classification.

While quantitative data were used in this report as one of the criteria for the type classification of the various type s of hyperlipoproteinemias, it is not suggested that these values represent any standard for other than illustrative purposes of this study. The data presented do indicate that quantitative classification offers potentially a more precise method than the subjective methods normally employed in cellulose acetate and agar gel interpretation and the ability to quantify the beta cholesterol or LDL obviates the need for an ultracentrifuge for accurate diagnosis. However, such information must still be used in conjunction with classical biochemical methods to accurately classify the lipoprotein patterns. A further advantage of using high resolution PAGE electrophoresis lies in the ability to study the lipoproteins in greater depth and to assess their relationship to other plasma proteins and enzymes.

Summary

A flat slab, step gradient, polyacrylamide gel system employing a discontinuous buffer system at a continuous pH is presented for the rapid separation and high resolution of pre-stained lipoproteins. Factors affecting resolution and quantitative methods to aid in the classification of the lipoprotein pattern types are discussed.

Acknowledgements

The author wishes to express his sincere appreciation to Dr. U. H. Klemens, Medizinische Klinik and Poliklinik of the Freien Universität Berlin, Klinikum Steglitz, and to Dr. W. Kahlke, Medizinische Universitätsklinik Heidelberg, for the generous gifts of sample material and helpful discussion. The author also wishes to thank Dr. H. Klein, Institut für gerichtliche Medizin, Universität Heidelberg, for the use of the ORTEC model 4300 microdensitometer and to acknowledge the expert technical assistance of Mrs. Rosemarie Hiemesch.

The author also wishes to express his sincere thanks and appreciation for his support by the Kultus-Ministerium and especially to acknowledge the friendship and support of Dr. Bernhard Urbaschek and Dr. Renate Urbaschek of the Institut für Hygiene und Med. Mikrobiologie Klinikum Mannheim der Universität Heidelberg who made this work possible during the author's tenure as a visiting professor in Heidelberg and Mannheim.

References

(1)   Moran, R. F., W. P. Castelli, and M. V. Moran, Clin. Chem. *18*, 217 (1972).
(2)   Raymond, S., J. L. Miles, and J. C. J. Lee, Science *151*, 346 (1966).
(3)   Wollenweber, S., and W. Kahlke, Clin. Chim. Acta. *29*, 411 (1970).
(4)   Maurer, H. R. and R. C. Allen, Z. Klin. Chem. u. klin. Biochem. *10*, 220 (1972).
(5)   Klemens, U. H.: Personal communication.
(6)   Ornstein, L.: Annals N. Y. Acad. Sci. *121*, 321 (1964).
(7)   Davis, B. J.: Annals N. Y. Acad. Sci. *121*, 404 (1964).
(8)   Canalco: Quick Disc Lipoprotein Diagnosis. (1969).

(9)   Allen, R. C., D. J. Moore, and R. H. Dilworth, J. Histochem. Cytochem. *17*, 189 (1969).

(10) Allen, R. C., This congress (1972).

(11) Allen, R. C., R. A. Popp, and D. J. Moore, J. Histochem. Cytochem. *13*, 249 (1965).

(12) Klemens, U. H., J. Schmalbeck, und E. Hendrich, Z. klin. Chem. u. klin. Biochem. *8*, 166 (1970).

(13) Fredrickson, D. S., R. I. Levy, and R. S. Lees, New Engl. J. Med. *276*, 32 94, 215, 273 (1967).

(14) Klemens, U. H., v. Lown of Menar, P., A. V. Bremer, E. Wnuck, und R. Schroder, Klin. Wschr. *50*, 139 (1972).

(15) Utermann, G.: Paper this congress (1972).

(16) Oette, K.: Der Krankenhaus-Arzt *8*, 439 (1972).

(17) Urbaschek, B., and R. C. Allen, unpublished data.

9.6 Human Serum Lipoproteins: Differentiation and Characterization by Disc Electrophoresis

G. Utermann

Human serum lipoproteins are found in a density range from 0.92 g/ml – 1.20 g/ml. According to minima and maxima of concentrations along a density gradient they are classified operationally into four major classes, termed Chylomicra, Very-Low-Density-Lipoproteins, Low-Density-Lipoproteins and High-Density-Lipoproteins (1,2,3). Within each density class lipoproteins are distributed as polydisperse systems varying in particle size and hydrated densities (4,5). Molecular weights of lipoproteins range from 170.000 daltons for HDL_3(6) to some million daltons for chylomicrons.

More recent work has demonstrated that the three main apolipoproteins Apo-A, -B, and -C are also distributed discontinously over the whole density range and a classification system based on these and connected findings has been proposed (7,8). The complex composition of ultracentrifugal lipoprotein fractions may be demonstrated by the fact that there have been isolated three classes of lipoproteins different in molecular weights, protein moieties and immunological and electrophoretical properties from the small density interval d=1.050 – 1.090 g/ml. Figure 1 shows the patterns of theses lipoproteins on polyacrylamide gel electrophoresis and summarizes some of the properties of the isolated fractions (8–16). This data indicate that there is no general disc-electrophoretic method available, allowing sufficient separation of all different kinds of lipoprotein molecules in a single gel concentration. This especially holds true, when looking at sera from patients suffering from disorders affecting lipoprotein metabolism. So this report can only depict some aspects and applications of the disc-electrophoretic method as applied to special problems.

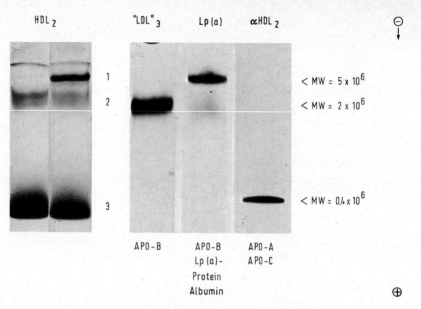

Fig. 1. Disc-electrophoresis on 3.75 % polyacrylamide-gels of two individual HDL$_2$-fractions
(left) and isolated lipoprotein classes from this density range (right).
Sudan black B-prestain.
For explanation and references see text.

When α1-lipoproteins of native sera are studied by disc-electrophoresis in a pH
8.9 system (17) using tris-glycine buffer pH 8.3 this results in a poor resolution
of diffuse bands. However, when 4×10^{-3} moles of lauric-acid are added to the
main gel a satisfactory and reproducible separation of different sharp α1-lipo-
protein components is achieved (18) (figure 2). Electrophoresis was performed
in a Shandon apparatus, using riboflavin catalyzed gels of 7.5 % monomer concentration. Thirty μl aliquotes of serum prestained with sudan-black B by the
method of Ressler et. al. (19) were applied to each gel. No sample gel is used. The
main components observed in native sera of apparently healthy persons were
designated α1 A, α1 B, α1 C and α1 D. The heavy band at the top of the main gel
represents LDL and VLDL; the faint band in front corresponds to albumin.

It should be stressed that it is of extreme importance for proper separation of
α1-lipoproteins to examine only fresh material. In aged sera new bands may
arise in the position between α1 A and albumin not observed previously in the
same sample. The nature of these bands is not yet clear, but there is an indication,
that they may arise from a disaggregation of VLDL (20).

When HDL subclasses HDL$_2$ (d = 1.063 $-$ 1.125 g/ml) and HDL$_3$ (d = 1.125 $-$
1.21 g/ml) are isolated under careful conditions (15), their α1-components exhibit

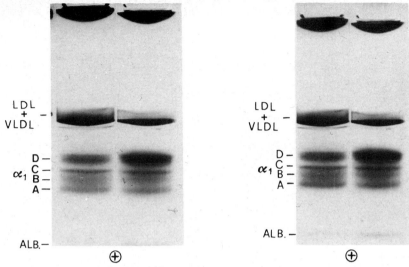

Fig. 2. Polyacrylamide-gel-electrophoresis of two individual sera in a 7.5 %, pH 8.9 system, containing lauric-acid in the separating gel.

different mobilities in the system. The HDL_3 fraction often shows two well-separated bands (figure 3). The HDL_3 fraction has a higher mobility than HDL_2-lipoproteins. HDL_2 exhibits a slow migrating $a1$-component. In addition, there is material in this fraction representing the high molecular weight lipoproteins containing apoprotein B, that do not penetrate the 7.5 % separating gel, but are separated on 3.75 % gels (figure 1). When the $a1$-HDL_2 and the HDL_3-bands were cut out and eluted from the gel, they were found to differ in apoprotein composition (21).

Comparison of HDL-fractions with the $a1$-lipoproteins of the original whole sera revealed that the $a1$-component of HDL_2 corresponds to band $a1$ D and that of HDL_3 to bands $a1$ A and $a1$ B. The correlation of $a1$ C is not quite clear, and it must be mentioned in this connection that the banding patterns of whole sera are always more complicated than that of the ultracentrifugally isolated fractions. This perhaps reflects changes induced by the isolation process in high salt concentrations.

Marked quantitative variations are observed in the intensity of the different $a1$-lipoprotein components between individual sera. According to their relative concentrations of $a1$-HDL_2 and $a1$-HDL_3 sera can be classified into three major groups (figure 3).

Sera containing predominantly HDL_3-lipoproteins (Type I), sera with HDL_2 and HDL_3 components in similar intensity (Type II), and sera containing predominantly the HDL_2 band (Type III).

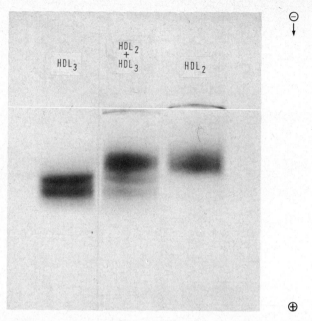

Fig. 3. Polyacrylamide-gel-electrophoresis of lipoprotein fractions HDL_2 (right) HDL_3 (left) and a mixture of both (middle).

Examples of the qualitative difference between individual sera are shown in samples 1–5 (figure 4). The differences in mobility between $a1$-HDL_2 and $a1$-HDL_3 reflect differences in the polypeptide composition. Thus, investigators may obtain quite different results, in respect to the quantitative distribution of Apolipoproteins, depending on the source of material.

According to recent findings of Alaupovic and co-workers (8, 22), lipoproteins containing Apo A- and lipoproteins containing Apo C-polypeptides are present as separate molecular entities in the HDL-fraction. An increase in concentration of one of the $a1$ HDL-bands does not, at present, provide any information on the protein moieties involved. So in each case, combined methods are needed to accurately define an individual band.

In the rare genetic disorder, lecithin-cholesterol-acyltransferase deficiency (LCAT-deficiency), abnormal lipoproteins are observed (23, 24, 25) which may also be demonstrated in biliary obstructed patients (26, 27). Figure 5 shows the comparison of the sera and HDL-fractions from a patient with LCAT-deficiency and from a normal blood doner using the lauric-acid-disc-electrophoretic method. No normal $a1$-lipoprotein bands are visible in the patient's serum. However, in his HDL_2-fraction we observed an Apo-A containing lipoprotein, that did not penetrate the 7.5 % main gel (26) behaving in this respect like B-lipoproteins.

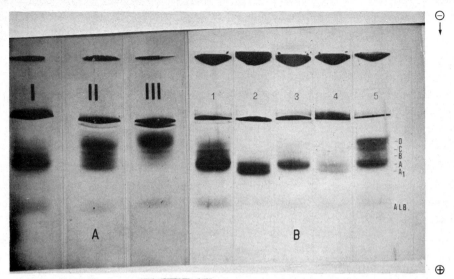

Fig. 4. Disc-electrophoretic a_1-lipoprotein patterns of individual sera, demonstrating the three main types I, II, and III (A). B: Disc-electrophoresis of five individual sudan black-B prestained sera. Sera 2 and 4 exhibit a fast moving additional a_1-lipoprotein component in the HDL$_3$ region (A_1). Alb.-Albumin.

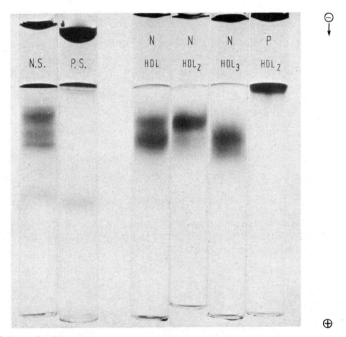

Fig. 5. Polyacrylamide gel-electrophoresis in the lauricacid system of a normal serum (NS) and serum from a patient with LCAT-deficiency (P.S.) and of isolated HDL-fractions from both persons (N=Normal; P=Patient with LCAT-deficiency).

One special advantage in studying serum lipoproteins is that prestained fractions react immunologically with the appropriate antisera against the native lipoproteins, thus allowing ready identification of electrophoretically separated components. Figure 6 shows the prestained LDL_2-fraction (d = 1.019 − 1.063 g/ml) from a patie suffering from LCAT-deficiency separated on a 3.75 % polyacrylamide-gel and subsequently studied by double diffusion in agarose. One LDL-component running in a normal position forms a prestained precipitate with an anti-LDL-serum, howeve no reaction occurs with the component on top of the main gel representing the abnormal lipoprotein-LP-X (28).

The application of these methods provides the possibility of characterizing the high density lipoproteins in greater depth and also as a technique to separate them for further biochemical analysis. It also allows an estimation of the relative HDL_2 and HDL_3 levels without resorting to prior ultracentrifugal methods, as well as aiding in the detection of possible genetic variants.

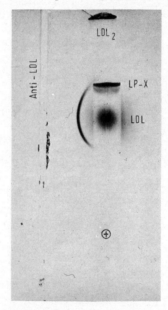

Fig. 6. Identification of a prestained lipoprotein band by combined disc-electrophoretical and immunological examination (for explanation see text).

References

(1)   Gofman, J. W., F. T. Lindgren, and H. Elliot, J. Biol. Chem. *179*, 973 (1949).
(2)   DeLalla, O. F. and J. W. Gofman, Methods Biochem. Anal. *1*, 459 (1954).
(3)   Lindgren, F. T., A. V. Nichols, T. L. Hayes, N. K. Freeman, and J. W. Gofman, Ann. N. Y. Acad. Sci. *72*, 826 (1959).

(4)   Oncley, J. L., Brain Lipids and Lipoproteins and the Leucodystrophies, Elsevier, New York, 1963.

(5)   Ewing, A. M., N. K. Freeman, and F. T. Lindgren, Advan. Lipid Res. 3, 25 (1965).

(6)   Hazelwood, R. N., J. Amer. chem. Soc. 80, 2152 (1958).

(7)   Alaupovic, P., in: Protides of the Biological Fluids ed. by H. Peeters 9–19, (1972).

(8)   Alaupovic, P. et al., Biochem. Biophys. Acta 260, 689 (1972).

(9)   Wiegandt, H., K. Lipp, und G. G. Wendt, Hoppe-Seyler's Z. Physiol. Chem. 349, 489 (1968).

(10) Utermann, G. und H. Wiegandt, Humangenetik 8, 39 (1969).

(11) Simons, K., Ch. Ehnholm, O. Renkonen, and B. Bloth, Acta path. microbiol. scand. Sect. B 78, 459 (1970).

(12) Ehnholm, C., H. Garoff, K. Simons, and H. Aro, Biochim. Biophys. Acta 236, 431 (1971).

(13) Albers, J. J., Ch-H. Chen, and F. Aladjem, Biochemistry 11, 57 (1972).

(14) Ehnholm, C., K. Simons, and H. Garoff, in: Protides of the Biological Fluids ed. by H. Peeters (1972), p. 191.

(15) Utermann, G., K. Lipp, und H. Wiegandt, Humangenetik 14, 142 (1972).

(16) Ehnholm, C., H. Garoff, O. Renkonen, and K. Simons, Biochemistry (in press).

(17) Maurer, H. R.,sDisk-Elektrophorese. Walter de Gruyter & Co., Berlin 1968.

(18) Utermann, G., Clin. Chim. Acta 36, 521 (1972).

(19) Ressler, N., R. Springgate, and J. Kaufman, J. Chromatog. 6, 409 (1961).

(20) Utermann, G., and H.-J. Menzel, unpublished results.

(21) Utermann, G., and H.-J. Menzel, Manuscript in preparation.

(22) Kostner, G., and P. Alaupovic, in: Protides of the Biological Fluids, (1972), p. 59, ed. by H. Peeters.

(23) Forte, T., K. R. Norum, J. A. Glomset, and A. V. Nichols, J. clin. Invest. 50, 1141 (1971).

(24) Norum, K. R., J. A. Glomset, A. V. Nichols, T. Forte, J. Clin. Invest. 50, 1131 (1971).

(25) Torsvik, H., M. H. Solass, and E. Gjone, Clin. Genet. 1, 139 (1970).

(26) Utermann, G., W. Schoenborn, K. H. Langer, and P. Dieker, Humangenetik 16, 295 (1972).

(27) Torsvik, H., K. Berg, H. N. Magnani, W. J. McConathy, and P. Alaupovic, Febs Letters 24, 165 (1972).

(28) Seidel, D., P. Alaupovic, and R. H. Furman, J. Clin. Invest. 48, 1211 (1966).

Abbreviations:

VLDL   = Very-Low-Density-Lipoproteins ($d < 1.006$ g/ml)
LDL   = Low-Density-Lipoproteins ($d = 1.006-1.063$ g/ml)
LDL_2 = „   „   „   ($d = 1.019-1.063$ g/ml)
HDL   = High-Density-Lipoproteins ($d = 1.063-1.21$ g/ml)
HDL_2 = „   „   „   ($d = 1.063-1.125$ g/ml)
HDL_3 = „   „   „   ($d = 1.125-1.21$ g/ml)

Summary of the Round Table Discussion and Concluding Remarks

R. C. Allen

Some eight years have passed since the first general conference on gel electro-phoresis. In that time we have seen a number of new techniques such as isoelec-tric focusing, SDS systems, new buffer systems in PAGE, and perhaps a bewil-dering array of equipment. Yet, there has been, if anythting, a proliferation of techniques rather than an attempt toward some standardization with which to compare results between laboratories and investigators.

Some of you have asked during the last two days why this meeting was so heavily weighted toward theoretical considerations rather than toward practical appli-cations and new methods. Both Dr. Maurer and myself agreed when we decided to convene this congress a year ago, that one of the basic requiremants prerequi-site to the possible standardization of techniques was a thorough understánding and airing of all aspects of the system in which we were working. This we believe is basic to comparing results between laboratories using past and present methods. I believe that Dr. Morris's physico-chemical studies, Dr. Richard's studies in polymerization kinetics, Dr. Staynov's work in non-aqueous systems, Dr. Jovin's developments of buffer theory and concepts, and Dr. Robards' and Dr. Chrambach theoretical considerations form much of the necessary groundwork from which to proceed in a logical fashion toward standardization of various systems with both continuous and discontinuous buffer systems. Dr. Vesterberg, Dr. Pogacar, and Dr. Catsimpoolas have, on the other hand, laid additional important groundwork in the area of isoelectric focusing.

Perhaps the question of polymerization catalysts poses one of the major problems to the success of many procedures. As Dr. Morris and Dr. Charambach have emphasized, the formation of peroxide is an intermediate in the polymerization of acrylamide, thus both persulfate and riboflavin catalization produce a resultant gel in a highly oxidized state. While pre-electrophoresis and washing in buffer can overcome this problem, it complicates procedures utilizing ionic strength and buffer ion discontinuities. Suggested procedures such as the addition of thioglycollate by Dr. Chrambach may help alleviate this problem, where pre-electrophoresis or washing of the gel is not possible and where an oxidized environment is undesirable for a specific application.

The observation that in many cases fomulas given in the literature will provide between 50 and 99.5 % polymerization have further complicated standardization. Certainly a 50 % polymerized 7.5 % gel will not behave similarly to one 99.5 % polymerized.

This brings up another point concerning the monomer itself. As produced commercially the crude acrylamide monomer contains a number of contaminents such as free acrylic acid, polyacrylic acid, B′ B″ B‴ nitrilotrispropionamide, free ammonia and ammonium bisulfate or calcium bisulfate depending on the manufacturer or origin.

Recrystallization of the monomer before use is prerequisite to standardization of procedures. However, as Dr. Hunter and myself found a number of years ago, monomer recrystallized from two sources using two different systems gave entirely different resolution in the albumin and pre-albumin esterase regions in the plasma of the same mouse.

I think that there is general agreement that gels from recrystallized monomer are tatamount to obtaining reliable and reproducible results not only between laboratories, but also within laboratories. Since recrystallized monomer will tend to rapidly deterioate on storage, storage under dry nitrogen below zero degrees centigrade appears desirable as is done with the commercially available 4 X recrystallized material from Serva Fine Biochemical. Similarly, methylene bisacrylamide should be recrystallized before use, and acetone appears from discussions here, to be the solvent of choice.

As Dr. Chrambach noted in the discussion, recrystallized materials polymerize with more difficulty than "dirty drude monomer and bis". Thus, we are faced with establishing some index oᴸ acceptable polymerization. While Dr. Morris has utilized 30 hours, others have stated 10—12 minutes is sufficient time. The appearance of a refractile line following water or buffer layering within 20 minutes seems to be generally agreed as being a reasonably satisfactory criterion when purified monomer is used in conjunction with appropriate amounts of catalyst and accelerator.

The use of prepared commercially available gradient gels has been discussed here by Dr. Hunter and has raised several pertinent points in the discussion. It appears that at present the stability of these gels and their reproducibility still remains to be proven. At present, we can only view their use with caution. However, as Dr. Jovin pointed out, industry should be encouraged to work more actively on this problem since such a material would certainly be valuable not only for large volume clinical use, but for the occasional user of electrophoresis

I would like to again mention the fascinating micro-gel methods that we have heard of from Drs. Grossbach, Neuhoff, Rüchel, Dames, Maurer and Giebel. The ability of these techniques to detect protein in the picogram range also provides a new and exciting potential.

In regard to preparative PAGE, considerable progress has been made in the last eight years with the theoretical development of the prep-opt system described by Dr. Chrambach and with the techniques and equipment described by Mr. Nees.

These, combined with the isotachophoretic techniques described by Drs. Griffith and Haglund, have broadened the potential use of PAGE as a preparative technique

The usefulness and potential of PAGE and PAGIF as clinical tools have been adequately demonstrated in these talks by Drs. Hoffmeister, Abraham, Felgenhauer, Uttermann and myself. It is studies such as these that are required to expand our vocabulary from the albumin, alpha-1, alpha-2, beta and gamma globulin era of plasma protein nomenclature to a broader and more meaningful one consisting of all the proteins presently identifiable by PAGE and PAGIF techniques. Not only is such an expansion necessary but it must be made in a manner both relevant and meaningful to our clinician colleagues. Also, in such a system the genetic information such as haptoglobin and GC type must be extracted first to avoid confusion. The present study of some 10,000 patient samples with up to 29 parameters by Drs. Hoffmeister and Abraham is a most significant step in this direction. Yet, at present, we have only reached a resolution capability of some 15–20 percent of the presumed number of proteins in plasma. As indicated by Dr. Chrambach's presentation on macromolecular mapping, we are perhaps approaching the resolution limit with discontinuous buffer systems in PAGE. On the other hand, isoelectric focusing can, as was demonstrated, detect 50 or more bands in animal plasma. Obviously, two dimensional techniques with both methods may extend our resolution limits and may provide a potential means to elucidate further protein species which may prove to have research or clinical significance. However, identification of each protein on such a map may prove to be a Herculean task.

Unfortunately, people in both clinical and research areas must wait for such information or do it themselves, which has and will prolong the interval from the present level of development of PAGE and PAGIF to their realizing their full potential as clinical tools.

The myriad of systems described here, both commercially available or homemade, horizontal, vertical flat slab and cylindrical, are so specialized and diverse as to cause the potential new user or neophyte considerable confusion to say the least.

This area certainly requires a greater degree of understanding directed at instrumentation with sufficient flexibility to perform PAGE, PAGIF, and immunological techniques in a single system or a system comprised of suitable components. One concensus that appears to come forth in the presentations and discussions in this regard is the desireability flat slab gels as opposed to cylinders. In this light, we should speak to a not wholly unexpected development in this congress.

In the last two days we seem to have identified, in addition to the numerous problems attendant with standardizing PAGE and PAGIF, several unexpected new finger colors. We have the blue fingers or the practical practitioners of electrophoresis, the red and black green, etc., associated with the theoretical considerations, isoenzymes, lipoproteins, etc., and finally, the gold fingers or those associated with the commercial aspects. Perhaps this somewhat facetious classification of the groups or sub-groups involved in this area of endeavor is one of the more important contributions of this congress. Certainly the problems to be resolved require the combined talents and close interaction of all of these groups. The theoretical considerations are paramount to extending PAGE and PAGIF to their fullest potential utilization. Yet, the practitioners of the art must determine the optimal conditions of practical application in research and clinical studies. All groups are to be a great degree dependent on that group Dr. Robards named the gold fingers, who provide the equipment reagents, etc. On the other hand, the gold fingers must be responsive to the needs and developments of the other groups independent of the immediate sales potential.

In a more serious vein, I think that it is evident to all sitting here that the free, and I trust unrestricted, exchange of information, ideas, and viewpoints over the last two days has been most beneficial and enlightening. There is, I believe, a clearer understanding of the constraints on industry, as well as their obligation in fulfilling the needs of the user.

On the other hand, I think there is a clearer recognition of the responsibility and role of those engaged in theoretical research and clinical applications to provide the appropriate direction for those who make and sell the equipment used by the majority in electrophoresis.

In closing, I would like to say that when Dr. Maurer and I planned this congress, we hoped that the word "standardization" in the title might stimulate a certain response in the participants. I don't think that we have been disappointed since we have certainly defined a variety of problem areas to be breached in the attainment of this goal. Until such problems are resolved, PAGE and PAGIF will remain, in the main, a research tool.

However, it would be presumptious at this point to attempt to set forth recommendations as to methods, procedures, or equipment, however, a number of cautions are clearly apparent from the various discussions, which I have tried to include in these closing remarks. I trust that the interest shown here in the last two days will be sufficiently lasting to provide some of these answers and to bring us together again in the not too distant future.

In behalf of Dr. Maurer and myself, I would like to thank all the speakers and participants for their excellent presentations and spirited discussions that followed both the papers and the Round Table.

I would also like to take this opportunity of again thanking the Erwin-Riesch-Stiftung and EMBO for their most generous support and similary thank those firms and exhibitors whose support helped make this congress possible.

Finally, I would like to thank on behalf of Dr. Maurer and myself, the Max-Planck-Institut for making these facilities available, and special thanks to Frau Maurer, Frau Pauldrach, her sister and brother-in-law for the success in the non-scientific evening events.

The editors would also like to express their sincere appreciation to Mrs. Martha Howard of the Pathology Department of the Medical University of South Carolina who performed to thankless task of typing and retyping many of the manuscripts for this congress.

Lecturers

K. Abraham
Institut für Sozialmedizin und Epidemiologie des Bundesgesundheitsamtes, 1 Berlin 33,
Postfach, Germany

R. C. Allen
Institut für Hygiene und Medizinische Mikrobiologie, Klinikum Mannheim
68 Mannheim D 6, 4–6, Germany
Present address: Department of Pathology, Medical University of South Carolina, Charleston,
S.C. 29401, USA

N. Catsimpoolas
Laboratory of Protein Chemistry, Central Soya Research Center,
1825 North Laramie, Chicago, Ill. 6039, USA
Present address: Massachusetts Institute of Technology, Dept. of Nutrition and Food
Science, Cambridge, Mass. 02139, USA

A. Chrambach
National Institute of Child Health and Human Development
Reproduction Research Branch, NIH, Bethesda, Md. 20014, USA

W. Dames, Max-Planck-Institut für Experimentelle Medizin, Abt. Neurochemie,
34 Göttingen, Hermann-Rein-Str. 3, Germany

K. Felgenhauer
Universitäts-Nervenklinik, 5 Köln 41, Joseph-Stelzmannstr. 9, Germany

W. Giebel
Hals-, Nasen- und Ohren-Klinik der Universität Tübingen,
74 Tübingen, Germany

D. Graesslin
Universitätsfrauenklinik, Hormonlabor, 2 Hamburg, Martinistr. 52, Germany

A. Griffith
P.O. Box 168, River Forest, Illinois 60305, USA

U. Grossbach
Max-Planck-Institut für Biologie, Abt. Beermann,
74 Tübingen, Spemannstr. 34, Germany

H. Haglund
LKB Producter AB, S-161 25 Bromma, Sweden

H. Hoffmeister
Institut für Sozialmedizin und Epidemiologie des Bundesgesundheitsamtes
1 Berlin 33, Postfach, Germany

R. L. Hunter
Department of Anatomy, University of California, Davis, Cal., USA

T. Jovin
Max-Planck-Institut für Physikalische Chemie, Abt. Mol. Biologie,
34 Göttingen, Postf. 968, Germany

G. Kapadia, National Institute of Child Health and Human Development, Reproduction
Research Branch,
Bethesda, Md. 20014, USA

O. Kling
Fa. Carl Zeiss, 7052 Oberkochen, Germany

H. R. Maurer
Max-Planck-Institut für Virusforschung, Abt. für Physikalische Biologie,
74 Tübingen, Spemannstr. 35, Germany

C. J. O. R. Morris
Department for Experimental Biochemistry, The London Hospital Medical College,
London E 1, England

St. Nees
Physiologisch-Chemisches Institut der Universität Bochum,
463 Bochum-Querenburg, Gebäude MA, Germany

V. Neuhoff
Max-Planck-Institut für experimentelle Medizin, Arbeitsgruppe Neurochemie,
34 Göttingen, Hermann-Reinstr. 3, Germany

G. R. Philipps
Physiologisch-chemisches Institut der Universität Bonn, 53 Bonn, Nußallee, Germany

P. Pogacar
Inst. für gerichtliche Medizin der Univ. Heidelberg
69 Heidelberg, Vosstr. 2, Germany

E. G. Richards
MRC Biophysics Research Unit. Department of Biophysics, King's College
26–29 Drury Lane, London WC 2, England

D. Rodbard
National Institute of Child Health and Human Development Reproduction Research
Branch, NIH, Bethesda, Md. 20014, USA

R. Rüchel
Max-Planck-Institut für experimentelle Medizin, Arbeitsgruppe Neurochemie,
34 Göttingen, Hermann-Reinstr. 3, Germany

H. Stegemann
Biologische Bundesanstalt, Institut für Biochemie,
33 Braunschweig, Messeweg 11, Germany

L. Strauch
Fa. Hoffmann La Roche AG, Abt. VI/Med., CH 4002 Basel,
Switzerland

J. Uriel
Institut de Recherches Scientifiques sur le Cancer,
Boite Postale No 894, 94 Villejuif, France

G. Utermann
Department of Serology and Bacteriology, University of Helsinki,
Haartmaninkatu 3, 00290 Helsinki, Finland

O. Vesterberg
National Institute Occupation, Safety and Health Occupation, Health Department,
Chemistry Department, Stockholm 34, Sweden

Subject Index